The Cloud-Climbing Railroad

N

La Luz elev. 4875

El Valle elev. 5370

Pinto elev. 6000

High Rolls elev. 6510

Mountain Park elev. 6710

Wooten elev. 7111

Toboggan elev. 7700

Baileys elev. 8100

Cloudcroft elev. 8600

Hotel Spur e. 8700

Cox Canyon elev. 8839

Russia elev. 9076

Alamogordo Jct. elev. 4402

Alamogordo elev. 4334

the
Alamogordo &
Sacramento Mtn.
Railway

The Cloud-Climbing Railroad

Highest Point on the Southern Pacific

By

DOROTHY JENSEN NEAL

ILLUSTRATED BY BOB STAGGS

Texas
Western
Press

THE CLOUD CLIMBING RAILROAD

Library of Congress Catalog Card Number: 98-85147

Consultant to first edition: Carl Hertzog

Cover Photograph: USDA Forest Service photo #233417

This Centennial Edition is dedicated to the memory of Dorothy Jensen Neal, and to her daughter, Judy Staggs, who made this edition possible through her donation of the copyright, and to her husband, Bob Staggs, who illustrated the book.

To Robert

My Husband and
Most Discriminating Critic

CONTENTS

PREFACE

TO THE 1998 CENTENNIAL EDITION

THE CLOUD-CLIMBING RAILROAD was first published more than 30 years ago, in 1966. At the time, Dorothy Jensen Neal little anticipated the high level of interest in an almost forgotten chapter in history. The first edition of the book sold out almost instantly, and the book went through several reprintings in the late 1960s. The book has now been out of print for several decades, but interest has remained high. In the intervening years, a number of individuals have continued to research the history of the Sacramento Mountain railroads, and new facts have come to light.

When the Sacramento Mountains Historical Society decided to sponsor a centennial edition of the book, a complete rewrite of the book was considered. However, it was concluded that changing the book would destroy Dorothy Neal's inimitable style and change the distinctive resonance of her writing. Thus, the book is being reprinted exactly as it was written in the original edition, and certain situations in the book, particularly as they pertain to individuals, remain frozen in the time frame of 1965.

No book has all the facts, and some "facts" do not always stand the test of time and further research. Where statements in the book require revision or further explanation, annotations have been added to the original text and appear as footnotes, flagged **, on or next to the page on which they occur.

Dorothy Neal died in 1974, but we hope that her spirit and enthusiasm for the Sacramento Mountains will live on in this book.

Olaf Rasmussen
July, 1998

Alamogordo

CHAPTER ONE

JUNCTION IN THE DESERT

TIMBER . . . timber, timber, timber! A cry of desperation penetrated the air of a barren desert. Answering echoes of promise reverberated from forest-crowned mountains nearby.

In the central southeastern area of New Mexico Charles Bishop Eddy was constructing the El Paso and Northeastern Railroad* and needed cross-ties for his track. Timber in the majestic Sacramento Mountains was close at hand and could resolve his difficulty, but it was practically inaccessible. A railway into the mountains was necessary to transport logs to the main line.

Cognizant also of the enormous quantity of excellent timber in the Sacramento Mountains, Eddy could not close his eyes to the possibility of profit from other uses for lumber. The mines in New Mexico and Arizona needed timbers for their shafts and the demand for lumber was rapidly increasing in El Paso, Texas.

This phenomenal promoter of the El Paso and Northeastern Railroad, although a native of New York, had already spent several years developing Pat Garrett's plan for the Pecos Irrigation and Improvement Company. In that venture Eddy had convinced James John Hagerman, a wealthy mining man from Colorado, to invest his money and time in the Pecos Valley.

---

* Eddy sold the El Paso and Northeastern Railroad to Phelps Dodge in 1905. It became known as the El Paso and Southwestern Railroad. In 1924, the Southern Pacific Railroad Company bought the El Paso and Southwestern.

William Ashton Hawkins[1], a young lawyer practicing in Silver City, New Mexico, met and became attorney for Eddy, joining him in the village of Eddy in 1889. This village, now the city of Carlsbad, was part of Eddy's Pecos Valley Ranch. Even if the promoter had spent years in searching, he could never have found a better qualified lawyer than Hawkins, whose legal ability and moral integrity were above reproach. Two men could not have been more complementary. Eddy, the promoter, the dreamer; and Hawkins, the businessman, the realist, produced a combination long remembered throughout New Mexico. Hawkins and his wife lived in the first brick house in Eddy County. It is still standing, facing the courthouse in Carlsbad.

CHARLES BISHOP EDDY. *(Courtesy of Archaeology and Historical Society of Carlsbad, New Mexico Museum and of William A. Keleher)*

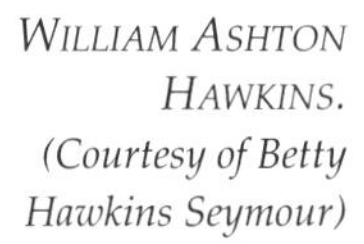

WILLIAM ASHTON HAWKINS. (Courtesy of Betty Hawkins Seymour)

The same year that Hawkins became associated with Eddy, William M. (Bud) Woods[2] joined the Pecos Valley project. Woods owned a ranch at Weed, New Mexico, located in the Sacramento Mountains. During the winter of 1888, a devastating snowstorm covered ranches in the Sacramento country with as much as eight feet of snow. Later, one rancher said, "That snow was deeper than a tall man's thigh on an awful tall horse."

Falling with no warning, the snow left many ranchers completely without their herds. Among them was Bud Woods. Many of the horses and cattle, along with deer and elk, were frozen to death as they stood in their tracks. In the spring

Woods received a letter from Charles Eddy informing him that some of his herd (branded 7-L) had drifted to the Pecos River. Eddy promised to hold the cattle in a pasture at the Eddy-Bissell headquarters. C. B. Eddy, his brother John A. Eddy, and their uncle, a Mr. Bissell, owned the Eddy-Bissell Cattle Company and used the brand VVN.

Immediately, Woods started his trip to regain his presumably lost herd. Possibly his gratitude and prompt response appealed to Eddy, for during Woods' stay Eddy confided to him plans for a town he hoped to build across the river. Eddy soon convinced Woods that he should make arrangements for his herd, move his family from Weed to the prospective town site and become foreman in charge of the improvements at Eddy, New Mexico.

By 1894, serious disagreements developed between C. B. Eddy and Hagerman over management of the Pecos Valley project. Differences of opinion grew into bitter altercations and within the year the Eddy brothers prepared to leave Eddy (now Carlsbad) the town which, like the county, bore their name.

Inspired by new fields to conquer, the dauntless Eddy began planning a railroad northeast from El Paso, Texas, into New Mexico. His plans included building a railroad village within one hundred miles of El Paso. Anxious to become acquainted with the territory he had heard about but not seen, he asked Woods if he had ever known of a spring called Alamo owned by Oliver M. Lee, a rancher. Woods knew the country and had once made a trip to the spring on horseback, but he was sure that Eddy could not travel to it by horse and buggy. Eddy ordered him to make ready a good team and buckboard and to prepare provisions for the trip.

When Woods announced the proposed trip to his family, his little son Carroll begged to go. Woods was adamant in his refusal, but on the following morning mentioned the boy's desire.

Eddy replied, "Bud, you let that boy go. I will never have a son of my own and I would enjoy having him along. He will see things that he will never forget. You and I will never forget them either."

On the trip they went through the Mescalero Indian Reservation where Woods made the mistake of letting the team graze on "sleepy grass." One of the needle grasses *(Stepa robusta),* it is found over much of the Southwest. It has a narcotic effect on horses and sheep but not on cattle. Ranchers say that it is more effective after a frost and that a horse, having eaten it, could be hit with a board, would jump ten feet then go right back to sleep. Carroll became furiously indignant when he realized that Indians were bringing their friends to sit on the hillside and laugh at his father's predicament.

From the Mescalero country they went to Tularosa which, at that time, was surrounded by an adobe wall. From that village the road to La Luz was a narrow wagon trail. There Woods received directions from former Texas cowboy friends, Jim and Millard Wayland, on how to cross the Red Arroyo and arrive at the mouth of Alamo Canyon. Since there was no road, they jostled along over gulleys, rocks and brush.

But, what a sight they beheld when they arrived at the place! There was a beautiful bowl-flowing spring surrounded by three huge cottonwood trees forming a triangle. These trees were more than five feet in diameter and shaded an area of over 150 feet. Nailed to one tree was a pine board lettered in axle grease with the words, "Ojo de Alamo Gordo." When Woods translated the words to Eddy as "Spring of the Big Cottonwood," Eddy exclaimed, "Alamo Gordo! Alamogordo! That will be the name of my town."

The story often told is that Eddy drank eagerly from the spring. Carroll says that the water was handed to him in a dipper for Eddy was too fastidious to lie down and drink from the ground.[3]

When the Eddy brothers left Carlsbad, they took Hawkins, Woods and others. Hawkins entered a law partnership in El Paso and for several years the family lived on Rio Grande Street in El Paso.

At once Eddy started laying plans for the railroad into northeastern New Mexico.** For two years he spent every waking hour investigating possibilities of coal mining in White Oaks and near Capitan, studying production of gold being mined at White Oaks, receiving reports on fruits being raised around Tularosa and La Luz, relishing statistics on herds of cattle on nearby ranches and delighting in the rumored extent and quality of timber in the Sacramento Mountains. He completely ignored the lack of good water and the dearth of timber in the desert site of his proposed railway.

Bolstering all his hopes and schemes was the legal advice he continually received from William A. Hawkins. Eddy, the charming persuader, and Hawkins, the correct performer, had thoroughly convinced Eastern capitalists and on October 28, 1897, incorporated the El Paso and Northeastern Railroad.

Eddy's first transaction was the purchase from George Jay Gould of ten miles of railroad already constructed northeast from El Paso. Gould had bought this unfinished project from Morris Locke who had intended to build it to White Oaks.[4] Immediately additional construction began with J. A. Eddy in charge. As in any construction crew, types of laborers varied greatly. There were at least fifty Chinese, each with a pigtail rolled up and held with a stick; several pugnacious Irishmen; about twenty well-built, hardworking Negroes; many Mexicans; some adventurous Texans and anyone owning, renting or borrowing a team and wagon. The men worked with scrapers, wheelbarrows, picks and shovels. Those driving teams were

** Localities in this paragraph are in southeastern—not northeastern—New Mexico.

proud of their ability to handle several animals at one time. A cook shack and tents kept close to the workers, moving every few days. Food from the commissary was good and prunes were plentiful. The men also bought shoes, gloves and overalls from the commissary and, as was customary, were always in debt.

Among employees that Eddy had previously hired was Jim Blakely. His reputation as a freighter was famous and well deserved. He drove a team of sixteen mules and named them all Pete. By the time the railroad reached Jarilla (Orogrande), Eddy had bought the Alamo Ranch and Spring for $5,000 and had started his railroad village of Alamogordo. Blakely hauled building supplies from the railroad to the town site over a road filled with ruts, holes, rocks and mud, caused from snow melting in the mountains. Often he cracked his long whip and shouted, "Feel of it, Pete? Don't you dare stick in this damn lob-lolly!" The very words must have cut for his team was never known to sink in the mire.

H. S. Church was the engineer for the Alamogordo Improvement Company, the organization building the town, and Bud Woods was in charge of surveying. Blakely hauled four wagon loads of one-by-four-inch Texas pine lumber that Woods ripped into two-inch strips and cut into stakes. Using a small compass and stepping off the distance, Woods set the stakes. Later, when Church corrected the survey, he found slight errors in direction but none in distance.

At the intersection of Tenth and Delaware Streets young Carroll asked why the streets did not go straight across each other. Church prophetically replied, "Son, some day this will slow down the traffic."

Among the surveyors was a man named Morton who was totally deaf. Carroll had a guitar and one evening Morton asked to borrow it. He then found a juniper stick about the size of a lead pencil. He placed one end of the stick against his teeth and touched the guitar where the neck joins the body with the other end of it. He tuned the guitar perfectly by feeling the pitch vibrations. Instead of strumming chords he played delightful melodies and plaintive tunes. Dick Poe, squatting with a cup of coffee, rubbed a tear off his cheek as he said, "Ain't that beautiful? I wisht he'd play that at my grave."[5]

Lots went on sale June 9, 1898, and Tom Frazier bought the first one. To it he moved a house from La Luz, the first wooden dwelling in Alamogordo. A tent village emerged immediately. Even the depot was a tent with a floor that was a twenty-by-fifty-foot frame made of two-by-eight-inch planks which was filled with sand. The north end of this tent served the railroad and the south end housed the Alamogordo Improvement Company.

Another dependable former employee Eddy had brought to Alamogordo was John Walker. His assignment was to bring irrigation water from La Luz Canyon to the new village. Because the mineral content of a glass of this water compared to a dose of Epsom salt, it was unfit for domestic use. Walker dug the ditch without help and proudly turned the water into it. When he arrived in town, a distance of five miles, so had the water. Realizing he must control the water, he turned his tired horse back toward the spring. By the time he cut off the water and drained the ditch, at least a foot of water had flooded all the tents of the village and completely saturated the sand floor of the depot.

When grading of the streets was completed, Woods supervised a crew of men who set out eight thousand cottonwood trees for future shade. W. D. Mayfield freighted the young trees from Las Cruces. Since they could not drink La Luz irrigation water, Carroll received twenty-five cents a day for carrying Alamo Spring water to his father's crew of about one hundred men. He remembers yet how, as he trudged along carrying his bucket, he hated any man who "rinsed out his mouth."

His brother, William Woods, Jr., aged seven, once expressed his opinion concerning temporary events. Mrs. Wornock, his teacher in Sunday School, asked, "William, can you tell us who made the world?"

ALAMOGORDO.
Alamogordo in 1902 at the intersection of Pennsylvania and Tenth Streets.
(Courtesy of Wiley Smith)

"Yes, Ma'am," responded the beaming youngster, "Mr. Eddy made the world, but my daddy planted the trees."[6]

Combined with beautification of the town were Eddy's plans for its inhabitants' welfare. Definitely opposed to the use of alcohol, he decided to limit its supply to his railroad employees. He resolved that there should be only one block in town where liquor could be sold. His intention was to have that block at the edge of town, away from the railroad offices. Hawkins, judicious as always, convinced him that it would be more practical to use a block in the center of town and within sight of the superintendent's office. It would take considerable thirst or little thought to warrant many trips to the bar and would, no doubt, completely prohibit some. One tenth of all liquor sales went toward building and maintaining a beautiful park which, parallel to the railroad, stretches a mile along the main street of the city even today. As yet, no attempt has ever reversed the restriction placed on location for the sale of liquor by Eddy and Hawkins within the original village limits.[7]

Alamo Spring furnished excellent drinking water for the town. A five-inch pipe conveyed water eight of the twelve miles to the depot. Gravity pressure of the flow was one hundred pounds per square inch. The first time a fire hose was attached, the pressure, until it was reduced, whipped the three men holding the hose wildly about the street.

A week after village lots were put on sale, at 10:00 a.m., June 15, 1898, the first train on the El Paso and Northeastern Railroad arrived in Alamogordo from El Paso with Henry Ackley as engineer.[8] During the months that followed he spent a great deal of time at the Baldwin Locomotive Works watching special locomotives being built for a railroad into the Sacramento Mountains which had to be constructed far enough to reach and transport timber for ties before the main line could continue northward. This Alamogordo and Sacramento Mountain Railroad was to join the El Paso and Northeastern at Alamogordo.

CHAPTER TWO

THE CORKSCREW ROUTE

CHARLES EDDY had started construction of the El Paso and Northeastern Railroad from El Paso to Alamogordo in October of 1897. Less than a month later he realized his shortage of timber for ties and organized the Alamogordo and Sacramento Mountain Railroad, a branch to be built into the Sacramentos. C. D. Simpson, his good friend, became president of the mountain line.

That Eddy was extremely busy and certainly well remembered in the Pecos Valley, the following telegram and note seem to prove. They were published in the *Eddy Current,*[1] by William H. Mullane. The telegram, dated December 10, 1897, was as follows: "On account of rumors which have reached here, I request to state that Hawkins' trip to see 'Gaullieur'* was about a matter that does not concern me. I have neither time nor desire to consider Pecos matters. Please assure friends." Mullane's remark followed: "Mr. Eddy probably has his hands full with his new railroad even though his interests in this valley approach a hundred thousand dollars."

---

* Henry Gaullieur[2], a Swiss government official, had recommended the Pecos Valley as the most favorable irrigation project in the United States to a group of wealthy Swiss immigrants who eventually settled there.

William A. Hawkins and his friend, Albert B. Fall, used the unwieldy size of Lincoln County for conducting ever increasing business affairs of the new El Paso and Northeastern Railroad to convince Governor Miguel Otero of the need for a new county.[3] Parts of Lincoln, Dona Ana and Socorro counties became Otero County on January 30, 1899, named in honor of the governor.

Not only Governor Otero but even laborers were proud of accomplishments and conditions in the area. One Irish carpenter hoped to share the opportunities offered him with a friend in the old country. He knew a conductor by the name of O'Shaughnessey.

Entering the station at Alamogordo one day, he asked, "Can I send a message to Ireland?"

"Of course," answered the agent.

"And how much will it cost?" he inquired.

The operator looked up the rate and said, "Write out your message and I will tell you."

He wrote: "Mike, come to America quick. O'Tero runs the government and O'Shaughnessey runs the railroad."[4] Indeed, Eddy's hands were full.

He had to build a railroad in order to build a railroad. Surveying for the mountain railroad from Alamogordo to Toboggan Gulch began in November, 1897. An affidavit signed by H. A. Sumner, Chief Engineer, shows a survey of the line of route made under his direction commencing November 10 and ending on September 1, 1898. This entire survey over both surveyed and unsurveyed lands was a distance of 19.34 miles. On October 20, 1898, the board of directors adopted his survey as the definite location of the railroad. In order to obtain benefits of the act of Congress approved March 3, 1875, entitled, "An act granting to railroads the right of way through public lands of the United States," a map was prepared of the survey. Also Sumner certified that the railroad was to be operated as a common carrier of passengers and freight.[5]

The El Paso and Northeastern, first named the El Paso and White Oaks Railway by Morris Locke, started on to Carrizozo from Alamogordo, and White Oaks eagerly awaited its arrival. Realizing the demand for coal in El Paso, from the main line Eddy decided to build a branch line to the Salado mines located near Capitan. He intended to add their supply to that of White Oaks, but—here we have an article from the *El Paso Times*[6] dated August 11, 1898:

> The White Oaks Railway will not go to White Oaks, but will turn off some distance this side to reach the Salado Coal fields 145 miles from El Paso. The 85 miles of the road already built brings us to the verge of the great timber belt and the first fifteen miles of the mountain branch now in course of construction will reach well within this belt. The mountain branch is known as the Alamogordo and Sacramento Mountain Railway. Five miles of the line is already completed to the old town of La Luz and the grading of the next ten miles is to be finished by September 1st. The mountain branch besides following one of the most picturesque routes in the world, will be a unique piece of engineering, with its exceedingly heavy grade and its sinuous windings up Fresnal Canyon.
>
> When proper accommodations are provided in the mountains El Paso business men will have what they have never had yet, a place where they can run out Saturday afternoons in summer, spend Sunday amid cool, picturesque and refreshing surroundings, and go back to the city Monday morning.

No doubt with this blasting of their hopes many White Oaks residents felt that Eddy deserved the blow when the Salado Coal fields were depleted within weeks after the branch line, which had been rushed to them, was finished.** But again

** The Salado coal fields were operated successfully for a number of years, although they did not live up to expectations. Production peaked at 171,000 tons in 1901 and declined thereafter, dropping to 1,898 tons by 1907.

Eddy proved his ability to cope with any situation. He envisioned unlimited profits from a timber business to be made possible by his little mountain railway.

His attorney, Hawkins, had filed certificates under general incorporation laws of the Territory of New Mexico on March 24, 1898, with $75,000 capital stock for the Alamogordo and Sacramento Mountain Railroad.[7] The charter authorized construction and operation of a railroad from Alamogordo into the timber belt of the Sacramento Mountains. Putting first things first, especially his need for ties to complete the main line, Eddy divided the construction crew and camp equipment in order to start building the mountain line.

Since blasting was necessary in the mountains, drillers and powdermen joined the crew. Construction foremen had warned native workers not to ride on wagons hauling explosives. Lon Hunter, son of the first elected sheriff of Otero County, and Wallace Beard were driving cattle from Peñasco Canyon down to Alamogordo when they met two Mexican laborers riding on a wagon pulled by six mules. Loaded in the wagon were rows of cans filled with black powder. The boys experienced difficulty in getting the cattle past the team and wagon. Its driver did not dare pull off the trail onto rough ground nor did he feel it necessary to stop and let the cattle pass. On their return trip, close to the "S" turn on the Fresnal, the boys encountered a gruesome spectacle. There were the remains of what had been six mules, a wagon and two men. The mules as well as the men had literally been blown to bits.

Mule teams were at a premium. John Fifer, father-in-law of E. P. McCrary, later a conductor on the mountain railroad, went to his mother's farm in East Texas near Goldthwaite to get mules. He used them for grading and hauling. Mules were more surefooted and less expensive to feed than horses. Until wagon trails were built, men carried equipment and material by hand.

Winter storms in the mountains retarded progress of construction and caused intense suffering for the crew. In spite of conditions, it is almost unbelievable the number of trestles or bridges the workers built on which they laid part of the track. There were 122 wooden boxes consisting of floor, sides and top averaging in size two feet by four feet with length averaging twenty feet. Fifty-eight pile or frame trestles averaged 120 feet long. The longest had two combined thirty-degree curves and a total length of 338 feet.

Length and height of each trestle depended on the width and depth of the canyon to be crossed. The rails were on cross-ties with a wooden guardrail on the outside to keep the ties from bunching. Under the ties were 8″ x 16″ x 28″** stringers put in three runs called cords under each rail. Cast-iron packers that resembled large spools spaced the stringers from two to six inches apart. Three-quarter-inch bolts held the packers and stringers together.

Bents were 12″ x 12″ upright timber posts 14 feet long set about 14 feet apart. The stringers joined and rested on their caps. The high panels were sway braced and set on intermediate caps about every 14 feet. Twelve-by-twelve-inch sills of necessary length held them. These sills were on footing blocks set underground from four to eight feet to resist danger of water.

Longitudinal braces ran through at each intermediate cap. Hog jaw braces extended diagonally from each intermediate cap to the next top cap at a forty-five-degree angle. They eliminated shifting endways or sideways. Sash braces ran across and sway braces, bolted on the side of the bents, ran down one side and reversed on the other. These braces helped keep lateral sway out of the trestle.

At the end of the trestle was a bulkhead which prevented dirt from sliding in on it. Made of 3″ x 12″ timber, the bulkhead was placed directly against the first bent on the dirt side.[8]

** This measurement should probably be 28′, not 28″.

S-BRIDGE—Note water barrels.
(Aultman Collection, El Paso Public Library)

S-BRIDGE.
(Late photo courtesy of S. A. Ramsdale)

TWO DENNY BOYS WITH CANT HOOKS, 1936.
Shows guardrails on all curves approaching trestles.
(Southern Pacific Collection)

Fireproofing protected the trestles from danger of fire caused by sparks as long as coal or wood was used for fuel. Gravel covered two-by-fours which ran between the ties blocking the gaps. Another precautionary measure was the use of guardrails on all trestles from La Luz to Cloudcroft, the village built by the railroad in the mountains. These extra rails kept a derailed train from jumping off a trestle.

Since all the lumber used was untreated, some of it needed replacement as often as every three years. A crew of at least thirty men was constantly busy changing ties and stringers and making other repairs. In replacing a timber on a trestle, the men used a push cart to haul it to the location on the track above, then lowered it with a heavy rope to the place where it was needed.

Curves as well as trestles made possible the trail into the clouds. With grades often of six percent and scenery of intrinsic beauty, its route afforded thrilling episodes and fascinating points of interest.

For approximately three miles the railroad ran northward without a bend over the first of the wood boxes and four pile trestles, three fourteen feet and one fifty-five feet long. Barely hinting of things to come were two one-degree curves. Over two more trestles and the line reached La Luz, 4.8 miles from Alamogordo Junction.

Turning southeast into La Luz Canyon, the road began to zigzag in earnest. During the first mile and a half only twenty-six-degree curves appeared, but just preceding a ninety-four-foot frame trestle were two curves of thirty-six degrees each, the greatest curvature on the line. (The angle which subtends an arc of 100 feet in length measures the degree of curvature.)

It was at the fork where La Luz and Fresnal Canyons joined that Roland Hazard, enthralled by the location, built a pottery plant in 1930. Using native clay and local workers, he produced ceramics that were exceptionally beautiful. They included tiles

RUSSIA—The end of the Alamogordo and Sacramento Mountain Railroad.
(Courtesy of Fred O. Bonnell)
Note: this photo appears to be Hubbell Springs in upper Hubbell Canyon. The photograph probably shows a Southwest Lumber Company camp ca. 1928–29. The track in the background leads to Marcia, the track to the left is the SWLCo main line to Lightning Lake and Agua Chiquita Canyon, and the track circling to the right in the foreground is a log spur leading northeast along the east slope of Hubbell Canyon.

ENGINE #2507 at Bailey's with log train, 1936.
(E. Clack Collection)

HIGHWAY CROSSING #1.
Near High Rolls, New Mexico, April 4, 1929.
(Southern Pacific Collection)

BAILEY CANYON TRESTLE.
Alamogordo and Sacramento Mountain Railroad,
north of Cloudcroft, New Mexico.

for roofs and floors, pots, and "strawberry" urns six and seven feet high. The products reached markets in many parts of the world. It was possible for passengers from the train to walk across the canyon and up to the pottery plant to make purchases.

Winding and twisting into Fresnal Canyon, the railroad reached El Valle, the first station and water tank in the mountains.** There it left the Fresnal and made its first 4.2% grade. At one point on a double horseshoe curve it was possible to see four trains at different places on the same track. Bridal Veil Falls, a cascade descending into a canyon forty feet below, dazzled viewers with its splendor.

The railroad ran almost straight northeast to Pinto, but with the help of two trestles and thirty-degree curves it made a hairpin curve changing its direction to southwest. This U-turn was so sharp that it enabled passengers in the cars to see the engineer. Eight thirty-degree curves in succession plus two long trestles made the trip from Pinto especially thrilling.

Fourteen and four-tenths miles was the distance from Alamogordo to High Rolls. Here was one of the fruit growing areas with which Eddy had enticed capitalists. Residents often brought fruit and flowers to sell or give to passengers at the depot. Others, as the train slowly traveled through their orchards, with aprons full of fruit tossed it through the open windows.

It was not unusual for deer to feed in the orchards at High Rolls. Mr. Mac (Conductor E. P. McCrary) and Jess Holden once shot a large buck there after hunting season had closed. They loaded it onto a log car and covered it with snow to refrigerate as well as conceal it. Imagine their surprise when a game warden boarded the caboose and rode to Alamogordo. For the remainder of the trip they spent their time either praying that the

** El Valle did not have a water tank. The only water tank between Alamogordo and Cloudcroft was at Wooten.

ENGINE #2506. *Lower leg of switchback with middle level trestle in background, 1936. (Collection of E. Clack)*

snow would not melt or planning a credible excuse for their violation. The officer left the train without even glancing ahead at the log cars.[9]

The shipping of ore as well as mining produced employment for the little village. The Wornock Copper Mine was west of Cloudcroft, the railroad resort built in the clouds, but this company shipped its ore from High Rolls. Wagons hauled it through the canyon, then the railroad delivered it to Alamogordo. From there the ore went to El Paso on the main line.[10]

H. W. Fleming of Cleveland, Ohio, organized the Standard Lithograph Stone Company, at High Rolls, in 1904, with capital of $15,000. On January 2, 1905, the railroad company furnished material and built a spur with labor and grading charged to the stone company.[11] Credit was given on freight shipments for the amount of labor and grading. During World War I, William G. McAdoo, Administrator of American Railroads, established this

THE SWITCHBACK—Lower right is Toboggan. Train is backing after coming down from Cloudcroft beyond upper right trestle, 1903.

basis for refunding to industries the cost of building part of their track. It is carried forward to this day as General Order 15—Supplement One.[12] The railroad had a siding 929 feet long in High Rolls.

For a very short time in the early 1920s Ben Longwell and his partner, C. M. Pate, shipped copper from High Rolls. Ben, still living in Cloudcroft, arrived with his family in Alamogordo as a boy of fifteen. He started working as a cook for the logging operations commenced in 1899 east and north of Toboggan. Later he worked as a forest ranger; but, by 1921, he let Pate persuade him that they should go into the mining business. Neither had any mining experience and Longwell started operating the mine alone, since Pate had business affairs elsewhere. Copper was then bringing twelve or thirteen cents a pound and costing Ben twenty-five cents to produce it by hand picking. Finally Longwell wrote to his friend: "Dear Pate, I'm trading off my part of the mine. May have to take a gallon of bootleg whiskey for it. Ben."

The answer came: "Dear Ben, Get another gallon and trade mine off too. Pate."[13]

Living today at High Rolls are Wiley Smith and his wife. For thirty years, as section foreman, Smith guarded the track of the mountain railroad. One morning after Hallowe'en he started on a hunting trip. Shortly after he left, Mrs. Smith saw a large log on the track that had been placed there by some pranksters. Leaving her year-old baby alone she ran down to the Post Office, a distance of about a half-mile, to inform her husband. Completely exhausted, she dragged herself back up the hill to her child. Smith and Damacio Nogales rolled the log off the track just as the oncoming train reached the spot.

Throughout the years Mrs. Smith was always ready to give assistance when needed by her husband or other railroad personnel. Once each year officials inspected the trestles. Her canned meat, vegetables and fruit were never more of a blessing than they were the day that Wiley was taking H. E. Stansberry, division engineer; F. M. Clough, superintendent of water and fuel and seven men from the San Francisco offices over the track. When they returned to the section house where the Smiths were living, she prepared a delicious meal for them in about thirty minutes. While they were eating, Mr. Clough noticed the red building paper, which covered walls and ceiling of the house, coming loose and billowing with every draft. He asked her, "What would you do if this place caught fire?"

"I'd grab my baby and run," she replied.

Soon afterward, Clough sent six-inch flooring for Smith to use to ceil the entire house.

High Rolls citizens once launched a campaign to attract health seekers to their village. Simon Kotosky, the much loved and respected owner of the general store, had his own idea for enticing passengers of the trains with merits of the village and his groceries. He insisted that his wife, Mrs. Beck and Mrs. Smith, all becomingly plump matrons, stand in front of his store as the train passed it going through town.

Around more thirty-degree curves, seven-tenths of a mile east, lay the village of Mountain Park. Today it and High Rolls are considered one community. With pride its residents claim a famous native son, cartoonist-author Bill Mauldin who often visited in the Smith home. Mrs. Smith remembers that Mauldin's sense of humor was apparent quite early in his childhood. Just south of Mountain Park, on the road to the cemetery was a highway bridge across the Fresnal. A plank broke and fell out near the middle of the bridge. Bill would wait under the bridge for an approaching car or pickup, then he would stick his head up through the gap. By the time the terrified driver was able to stop his vehicle on the slope off the bridge, the boy had ducked down and was hiding in some brush nearby.

At Mountain Park, more orchards produced the finest of apples, cherries, pears and peaches. In 1907, S. F. Cadwallader and Sons had 30,000 fruit trees and a nursery with many shrubs and plants. They also built a warehouse ninety-six feet long and twenty feet wide. From the nursery, every Friday afternoon, Engineer Ernest Clack** took one dozen gladiolas home to his wife in El Paso. These flowers lasted until the following week.

One attempt at raising cauliflower at Mountain Park, although an abundant and superior crop, was a disappointment. The heads turned yellow before they could reach a market.

The *El Paso Times*[14] of September 29, 1925, states: "The Mountain Park, New Mexico, area will ship 120 cars of apples this year, according to Richard Warren of the Southern Pacific Railroad Company."

** Ernest Clack was a conductor rather than an engineer.

ALAMOGORDO LUMBER CO. SHAY #3—Running as helper on the S-Bridge, 1899.
(Taken from "The Cloud Climbing Route" published by
Texas and Pacific Railroad Co. in 1900.)

SHAY ON MEXICAN CANYON BRIDGE.
(Taken from "The Cloud Climbing Route" published by
Texas and Pacific Railroad Co. in 1900.)

The railroad installed a three-car spur at Mountain Park, in 1908, which cost $434.43. In 1909, the company extended the spur using old track from Toboggan farther up the line. McRae Lumber Company did grading and furnished cross-ties. Other railroad properties at Mountain Park were an irrigation syphon enclosed in a frost box and a pergola station building.

Leaving Mountain Park and crossing the Alex and H. B. Harris property lines, the railroad reached the B. F. Wooten property and Wooten Station. Mrs. Wooten's sister, Mrs. Ida Wofford, now ninety-nine years of age, lives in Alamogordo. She still proudly displays scrapbooks she has kept throughout the years of interesting events on the Alamogordo and Sacramento Mountain Railroad.

Wootens owned the orchard and ranch from which the railroad received water. A spring fed a tank sixteen feet in diameter which held water in storage. In 1906, an analysis of Wooten water showed it was "bad" having 4.57 pounds per 1000 gallons of encrusting solids. About ten years later records show that the company furnished a treater, raised the Wooten water tank on account of height of tanks on big Shays and rod engines, replaced the four-inch pipeline from the spring, put a substructure under the treater and started using wooden instead of cheaper iron tanks. Even so, undesirable solids encrusted locomotive boilers until the company replaced all treater tanks in 1924.[15]

Too much uncontrolled water, on the other hand, could be as disastrous as not enough good water. At several points along the line, open chutes collected water above the track and aimed it over the track, thereby preventing washouts. One of these chutes was twenty feet high.

Climbing and curving from Wooten the railroad reached a canyon both narrow and steep. Since it took time as well as technique to cross it and climb out, the lumber company built the first logging camp at the bottom of the canyon and named it Toboggan. It was there also that the construction crew camped while awaiting further instructions. Building the camp was scarcely completed when a waterspout, flooding the canyon, gave only enough warning for the inhabitants to leave their shacks and run up the hillside. Horses, mules, oxen, harness and other equipment washed down the canyon to La Luz.

The problem of crossing the canyon at Toboggan and climbing the mountain was solved by a switchback, one of the most famous feats of engineering on the railroad. Legend says that the steepness of the mountain had completely baffled the original engineer and that his son gave him the idea for the switchback. The age of the boy has varied from youngster to college student as has also the mental health of the father. Versions differ, one reporting the boy to have been playing with toys when the father, discussing the frustrating problem with his wife, jokingly asked, "Son, what would you do?"

The boy replied, "See, Papa. I'd go up here, then come back here and start back up," as he showed his father with the blocks he had arranged on the floor.

Then the father, clasping both son and wife in his arms, shouted, "That's it! That's it! Mary, he has saved us!"

Less dramatic, indeed, is the story that worry over the complication had caused the father to have a complete nervous breakdown and that his son, studying engineering, had come home from college with the idea for the switchback.

This marvel was built so that the train traveled northeast to the end of the track. Then a switch let it travel backward onto the switchback track. For more than a mile northwest, over two trestles, of 108 and 200 feet respectively, and a curve of 22 degrees, the train backed to the north end of the track. From there the engine headed out, switching onto the upper track going east toward Cloudcroft. For some distance the track ran parallel to the lower track. The top of the switchback could hold only twenty-two cars without backing them into the mountain. There was a grade of six percent at the bottom of the switchback which made the return trip often very difficult.

Going either up or down the mountain line the train always went backward on the switchback.

On the 169-foot trestle just above the switchback Fireman B. B. Wilson of El Paso encountered his most frightening experience. He was on the way down to Alamogordo with Engineer Bill Reynolds who was making his first trip on the mountain road. Reynolds was an experienced mountain engineer from Colorado. As they started onto the bridge with a load of logs, a tank wheel broke into pieces resembling cut portions of a pie. The oil-line broke, making it impossible to get fuel into the firebox (locomotives were burning oil at the time). They called a section crew who cut up cross-ties and loaded them into the rack. Wilson fired the remainder of the way into Alamogordo with wood.

Before reaching Bailey's Canyon most of the grades were over five and some nearly six percent. Here was the big horseshoe curve, first a 30-degree curve with a 198-foot trestle, then a 261-foot trestle on a 28-degree curve. The swaying and creaking of the trestles often made the passengers think that the bridges were falling to pieces. A spur laid in 1911 had been taken up in 1912 because it was too steep. However, in 1920, the company completed a 350-foot side track which, according to records, cost $927.00.

It was on this horseshoe curve at Bailey's Canyon that a car once became completely detached from a moving train. On this train was Mrs. Jack Neece, wife of the tie inspector, who often accompanied her husband to Cloudcroft where she sold silk hose. After a day in the mountains, she and their children were returning to Alamogordo riding in the caboose. Next to it was a boxcar loaded with cord wood. When the train reached the big curve, the boxcar came loose from the car ahead. Mrs. Neese said she prayed, "God, if Thee ever helped us, help us now." Just then the boxcar uncoupled from the caboose, went three hundred feet to the bottom of the canyon and turned over. Sitting in the cupola, the horrified conductor could only mutter, "This is it, this is it." Then the caboose, still shuddering and weaving dangerously, began to pick up speed going down hill. The rest of the train had slowed down going up the next grade and with a jolt the caboose hit and they hooked back together. The brakeman then connected the air hose and the train went on to Alamogordo. The boxcar was never recovered.[16]

Next was a 323-foot trestle called the Mexican Canyon Trestle. From it the train whistle alerted the town of Cloudcroft but it was at least fifteen minutes before the train reached the depot. Residents of the village could determine the location of the train by the sound of its wheels on the curves. A whistling noise, almost a chord, varied slightly on each curve of the track.

Just beyond the Mexican Canyon Trestle was the sixty-foot high "S" curve bridge**, another wonder of the climbing railroad. It was a 338-foot frame trestle on reverse right and left curves. Truly an inspiration to many photographers, details of

---

** Using the photograph on page 16 and standard railroad measurements, the S-bridge was probably about 41 feet, 4 inches high: 2 posts, each 14 feet in length, plus additional posts of varying heights less than 14 feet long (probably about 10 feet) to conform to the canyon floor; a 12" cap; a 16" stringer; an 8" tie; a 4" guard timber.

the bridge and trees in the background have remained unchanged for nearly half a century.

Especially inviting to children was this bridge. On it one day three youngsters were sauntering along, throwing pebbles over the edge. Not one of them heard an approaching train until it was too late to get off the trestle. They all jumped into the water barrel on the edge of the trestle. When the train passed, they emerged thoroughly soaked but safe.

As a boy, Raymond Buckner, now of Ruidoso, had a small wagon. He would fit the wheels of one side of it on the metal track and those of the other side on the flat wooden rail on the edge of the "S" trestle. He always leaned to the inside as he rode until his parents learned of the sport. It is not hard to imagine that he then assumed another angle.

Dee T. McLean, however, was firing one morning on a train traveling over the "S" bridge; the floor of the engine cab was wet and slippery causing him to lose his footing, and out he fell over the edge of the bridge but managed to grasp the bottom step of the engine with his fingertips. There he dangled, hanging on for dear life, until the locomotive got to the end of the trestle where he could jump off onto the ground.

From the"'S" bridge the railroad ran southeast and east to Cloudcroft. More curves up to thirty degrees preceded a spur of 231 feet just before a 124-foot frame trestle. The horseshoe curve as the road entered the mountain resort caused several accidents. The late George Ramsdale, a brakeman and extra conductor, one evening had set the brakes on a log train at that point. He got off the train to throw the switch. Ice, which had formed on it, caused him to spend more time than he had anticipated. Suddenly he realized that the brakes on the cars had let go and that the train was passing him. His next sensation was that of hearing the log cars pile up on the depot platform. After working all night to help remedy the situation, he arrived home in Alamogordo.

At the door he announced, "Grace, I've played hell."

"I'm not surprised. You've had all night to do it," snapped his anxious wife, aware only that he had been gone all night.

When told of the catastrophe, thinking that he would be fired, she decided that they should start to pack their belongings. George persuaded her to wait for a day while he made a trip to the offices in El Paso. From there he returned to his job with a handful of "brownies" (demerits).

Leaving the town on twenty-six and thirty-degree curves, the railroad went east to a 154-foot trestle making another U-turn of thirty degrees. Then the road headed southeast to Cox Canyon. Here was a 625-foot spur once holding a number of bunk cars. Winter was already at hand, on November 16, 1919, when seven collapsible bunkhouses for section laborers were ordered for Cox Canyon from El Paso. They arrived January 30 with no bolts to hold on the roofs. By February 10, an order sent 800 redwood ties with which to start construction on a building to replace the collapsible bunkhouses.[17]

For about two miles over more curves and trestles the line traveled generally southwest then south to Hudman's Spur. In 1917, the railroad had built a 250-foot spur to accommodate E. E. Hudman when loading wood. More trestles, including one 176 feet in length completed the line to Russia Station, so named because of its extremely severe winters.

It was at Russia Station where Mr. and Mrs. Wiley Smith were living in an outfit car when their daughter Fern was born. Coincidentally Dr. G. W. R. Smith from Cloudcroft was the attending physician. Mrs. Smith had a nurse by the name of Smith. Wiley's children, by a former marriage, and the new baby just added more Smiths to the carload.

At Russia, the end of the line, was a wye with a thirty-degree curve. A track at the beginning of the wye led to the Alamogordo Lumber Company logging railroad. The end of the wye was 31.12341 miles from Alamogordo and marked the end of Eddy's Alamogordo and Sacramento Mountain Railroad.

CHAPTER THREE

FROM CANYON TO CLOUDS

THE FIRST STOP on the Alamogordo and Sacramento Mountain Railroad after leaving the town of Alamogordo was the picturesque adobe village of La Luz. When searching for a date of a possible early settlement, W. A. Hawkins found a record of a baptism in the area in 1719. On a bell he had forged for his home, La Claridad, built in La Luz, he used that date. No other record of a village until after the middle of the following century is available.

Early in the 1860s, a caravan of settlers, natives of towns along the Rio Grande from Albuquerque south to Socorro, trekked to the west edge of the Sacramento Mountains to take up homesteads. Fertile soil and mild climate awaited them at the entrance of the canyon. This seemingly terrestrial paradise, however, suffered from continual harassment of marauding Indians.

Traveling by wagon train the migrants had stopped in Tularosa before the final stage of their journey. Several fascinating stories exist as to the naming of La Luz. The most plausible one is related here. The travelers had promised their new friends, whose wall helped protect them from the Indians, that after they arrived at their destination, each evening they would build a huge fire. When this was done its light shining across the desert signaled to the anxious watchers in Tularosa that all was well. No doubt they said, "Allá está la luz. Está bien."

For the first few days the newcomers combined their efforts in building "La Muralla." This wall enclosed an area large enough to protect the entire group. The home of Mrs. C. H. Haynes now is on the site of the area used to safeguard the animals. The old fortification is now Presidio Park.

José Manuel Gutiérrez was one of the leaders of the group. He was born at Lemitar, New Mexico, in 1806 and died at La Luz in 1889. His son was José Dolores Gutiérrez who was the father of Juan José Gutiérrez living in La Luz today. Juan José was born at the mouth of the canyon where his father had homesteaded 160 acres. He worked in Alamogordo as a railroad mechanic for thirty years. He and his wife are justifiably proud of a family of seven sons. They are Albert B., employed by Douglas Aircraft, Long Beach, California; Ernest B., American Consul, Cairo, Egypt; Steve and Richard, Holloman Air Force Base; Fidel, teacher at Irvin High School, El Paso; Rudolfo, employed by El Paso Hotel Supply and Gilbert, a disabled veteran working on property at La Luz. Ernest fondly remembers free rides they all enjoyed on the mountain railroad.

Another leader of the band of settlers was Bartolo Pino. With him came his wife and three stepsons, Maximo, Manuel and Juan García. The boys ranged from twelve to sixteen years of age when they arrived at La Luz. Juan's son, Juan García, Jr., and his wife are residents of La Luz today. In 1957 their daughter Bertha (Mrs. Joe Carabajal) was first queen of the annual fiesta.

Stories told by Juan García, Sr., to his family concerned Indians. At one time Bartolo Pino, Juan's stepfather, chased a band of over 500 Indians on a hill up La Luz Canyon with only one bullet in his gun. Waving the gun and shouting for help, he put the Indians to flight. Although hit in the hip by an arrow, he pulled it out and kept on screaming, "You did not hurt me!"

Another Indian story from the García family bears repeating. Four of the men who brought cattle along with the wagon train hoped to establish and develop ranches. Without warning, one morning, they found themselves helpless against a band of many Indians who drove their precious herd of forty into the mountains. The ultimate outrage came when the settlers saw the plunderers butcher their cows, pack the meat on their horses and ride away.

The Baca brothers, Martín, Antonio, and Francisco were among the courageous pioneers. Francisco was the grandfather of Mrs. Juan José Gutierrez. He was killed by Indians in 1876. His house is now the Crocket Store. Martín's home was next door to the north. Antonio built the house which is now part of The Inn (Mi Casa).

Other founders of La Luz were Bentura Girón, Seferino, Luciano and Felipe Gallegos, José Moya, José Barela and Lorenzo Ulivari. Ulivari owned a large tract of land extending from the west edge of the village eastward to the site of his house. Near its location now stands the old railroad section house. He planted his land in wheat, harvested it by hand and hauled it to the Ostic Mill.

William Ostic, who built a flour mill up La Luz Canyon, William Gentry and Perry Kearney were among other early settlers. The population of La Luz was 249 in 1880. By this time Kearney had cultivated a large orchard and in 1883 had received homestead papers. He then divided his land into tracts to sell to others. With shrewd foresight he assigned irrigation rights to each acreage or lot.

These early settlers bought and traded land without surveying it or recording deeds. From a certain rock to a designated tree marked a boundary and a man's word meant as much as his signature. A common arrangement for mutual advantage was to trade a horse and saddle or a team for a piece of land.

La Luz

When the railroad started building into the mountains, many of the farmers in the canyon sold their land and water rights to the company for water in its village of Alamogordo. La Luz Canyon water was used only for irrigation in the railroad town because of its high mineral content. Since the canyon farms were without water, they were of no value to anyone. Many of the former owners bought acreages in La Luz using remaining water rights from an open ditch and much later developing their own wells.

A Mr. Smith became noted for his fine vegetables and delicious fruit raised in the village. Orris Smith of Capitan remembers moving into a building next door to him in 1904. Since the winters were most severe in the Sacramento Mountains where she lived, Mrs. W. L. Smith, mother of Orris, decided to move with her children into La Luz. They rode the Alamogordo Lumber Company Shay #2 up from the Peñasco. The engine crew, carrying the boys, broke the trail through snow four feet deep to enable the family to catch the main line train #185 and ride into town. Children of Charlie Myers, the grocer, were playmates of the Smith boys. Three days a week Elzie Swift met the train with his horse and buggy and delivered mail to the post office. The railroad station was about a mile from the center of the village.

Completely charmed by the little town, W. A. Hawkins built a home in La Luz. For years while the family lived in El Paso, this dedicated man, involved in his voluminous legal business, was almost a visitor in his own household. His seven children, three girls and four boys, held him in awe, fascinated by his tales, but rather dismayed by his stern and dominating manner. Mrs. Hawkins, withdrawn and nearly overshadowed by her enterprising husband, held fast to Continental European customs which she had inherited in New York. Names of eastern garden flowers rather than of western cacti she made familiar to her children. She felt eastern vacations were necessary to complete their family life.

TIE BUNKHOUSE for section workers.
Wheel assembly on automobiles rigged for use on tracks.
(Southern Pacific Collection)

INSIDE TIE BUNKHOUSE at La Luz.
(Southern Pacific Collection)

The girls attended El Paso School for Girls (Radford) and occasionally invited faculty members to the home for dinner. Their mother beamed with satisfaction if the guest used an eastern accent while the girls giggled behind their napkins.

Hawkins appeared constantly to be protecting his family from the frontier although he was completely engrossed in developing it. Even though his stories sounded romantic and his wife maintained that she had never heard the same one twice, he seldom brought frontier characters to the home. One exception was Miguel Otero, whose dynamic presence at a family dinner made it a momentous occasion.[1]

Mr. Hawkins undoubtedly was away from home the day in 1915 when the body of General Pascual Orozco lay in state in a house across the street from the family residence. Orozco[2] with four of his faithful companions had been killed near Van Horn, Texas. The bodies were packed in ice and put on a train for El Paso. Mayor Tom Lea, fearing violence from angry Mexican sympathizers, had Orozco's body removed from the train east of El Paso and secretly brought to a house[3] on Brown Street between Rio Grande and Montana Streets. Nearly all day, using the pretense of roller-skating, the Hawkins children watched; first, the body carried in on a stretcher, then the continuous stream of mourning and wailing figures who slowly entered and left the house.

For some years, during the summers, the Hawkins family spent time at the charming and commodious ranch of Albert B. Fall at Tres Ritos, New Mexico. By 1919, however, the family started spending summers at La Luz. There Hawkins and Charles Sutton remodeled The Inn, the old stagecoach stop of the 1880s, adding eight bedrooms with baths and a large dining room opening onto the luxuriant garden. It is now Mi Casa, owned by Mrs. A. T. Seymour (Betty Hawkins) of Albuquerque. Her son, Leonard Wood, and wife are restoring it as their private residence.

Cloudcroft

The Hawkins family moved into the spacious house the attorney built on the south edge of La Luz when he retired in 1927. It was here he used the bell with the date 1719. Bert Pfingsten of Hondo once asked him why he kept adding rooms to the place. Hawkins answered, "When you quit building, you are finished."

He built a golf course across the road from his house extending south and east, over and under the railroad. On it he had the only steel golf greens in the Americas, made from filings obtained from the railroad shops. Creosote bush and mesquite had to be cleared from the fairways. Tees were boxes filled with shale above the ground level. In order to see them on the beige-colored earth, players used red balls and tees. Attorney Ben Howell of El Paso remembers playing on this course in the 1920s. A standing joke was that the greens were of steel so that they would not blow away.

The attractive house, surrounded by huge trees and ornamental shrubs and enclosed by a low adobe wall, today retains all its splendor and clarity. Major and Mrs. Jacob Quintis are now proud owners of La Claridad de La Luz.

Entering the Sacramento Mountains from La Luz, H. A. Sumner and his surveying party reached Toboggan on September 1, 1898. Within the next few months they had penetrated territory heretofore viewed and described glowingly only by hunters and prospectors. The United States Government had withheld this land from settlement awaiting decision on its becoming a Forest Reservation. The remote locality, within one hundred miles of El Paso, was one of the most unexplored sections of the entire country.

On the summit of the mountain the surveyors rested. It was here they looked to the west and beyond the sheer descent could clearly see the White Sands in the distance. Eastward the country sloped gradually offering vast expanses of ever-changing forest that could easly be traversed. Inspired by the sight of fluffy clouds lying near the ground, they named the

MAIN STREET—Cloudcroft, New Mexico, about 1905.
(Aultman Collection, El Paso Public Library)

area Cloudcroft—a pasture for clouds. Here, they felt, was the site most suitable for a village summer resort and luxury hotel for the railroad company.

By spring of 1899, the railroad had arrived at Toboggan and construction of the village of Cloudcroft was under way.** The railroad company set aside 2700 acres, a tract of land on the summit approximately 2½ miles long and 2¼ miles wide, for the Cloudcroft reservation. Climate and beauty, now readily accessible, at once achieved unprecedented miracles through the medium of advertising.

** The railroad arrived in Toboggan in June of 1899 and in Cloudcroft in May of 1900. For about a year Cloudcroft passengers transferred from the train to horse-drawn wagons at Toboggan for the last part of their trip.

The first advertisement for the completed Alamogordo and Sacramento Mountain Railway in the *White Oaks Eagle*[4] ran June 15, 1899, as shown on the next page.

Charles Eddy and his railroad company were determined to provide the comforts of civilization for this heretofore uninhabited expanse of beauty. Located on the summit of the mountain, it had no natural streams. At the cost of $30,000 they constructed a water system from James Canyon which supplied the town. Pumped through a six-inch standard cast-iron water main, over an elevation of 625 feet, water was available to inhabitants of the entire community.

Beginning in 1922, water was pumped in three stages from James Canyon. From the pump house at Moser's Spring an eight-inch suction line conveyed water to a thirty-foot concrete tank. From this tank to a steel tank was a six-inch pipeline and from the steel tank a four-inch pipe carried water to the Lodge tank.[5]

Water service assisted the progress of constructing residences and other buildings. Until their completion, the Pavilion, built close to the depot where passenger trains stopped, provided for the many needs of visitors and future inhabitants. It was a building 170 feet long and 50 feet wide with kitchen and dining facilities for feeding hundreds of people. Included were bathrooms, parlor, reception rooms and porches on all sides. People camping or in unfinished cottages could obtain meals there until their residences or hotels were completed. Early arrivals lived in tents or rooms in cottages near the Pavilion. Rent was reasonable for these temporary accommodations. At once it became apparent that many families would delight in the wholesome recreation and relaxation of this wooded paradise.

The many forest animals and birds provided inspiration for names of streets. Curlew, Wren and Swallow Places wound the length of the town north and south; and Possum, Chipmunk, Squirrel, Coyote, Fox, Lynx, Panther, Fawn and Otter Avenues crossed them.

El Paso and Northeastern
and
Alamogordo and Sacramento Mountain Railways

Time Table No. 2
(Mountain Time)

Train No. 1 leaves El Paso . . . . .10:30 A.M.
Train No. 2 arrives El Paso . . . . .7:15 P.M.
(Daily except Sunday)

Connects at Alamogordo with stage line to Nogal, Mescalero, Ft. Stanton and White Oaks.

No one should leave Alamogordo without making a trip on the Alamogordo and Sacramento Mountain Railway.

That Famous
"Cloud Climbing Route"
and cool off
at
"Cloudcroft"
The breathing spot of the Southwest.

For information of any kind regarding the railroad or the country adjacent thereto, call on or write to:

A. J. Greig
Gen. Supt. & Gen. F.&P. Agy.
Alamogordo, New Mexico
or H. Alexander
Asst. G. F. & P. Agt.
Alamogordo, New Mexico
or F. E. Morris,
Local Agent
El Paso, Texas

EL PASO TIMES advertisement, August 31, 1907.
(Courtesy of John White)

Planning for a large resort, the company divided the Cloudcroft Reservation into three sections. North Cloudcroft was to be the commercial community. A street one hundred feet wide, through which a fence with gates was to be placed if necessary, separated it from the rest of the town. Lots there were sold without restriction except as against liquor. The company operated the only bar and insisted that the morals of the entire community be above reproach.

Cloudcroft, the hotel and compact portion, became town lots. They were 40 by 100 feet and were sold subject to restriction against liquor; trade or commercial purposes, except boardinghouses and hotels; and the keeping of horses, mules, burros, cows, swine and domestic fowls. Lots ranged in price from $50 to $200 with 50% more for corners.

Cloudcroft Park, the outlying section, consisted of acreages for those preferring seclusion or more land for large residences. These small tracts had restrictions only against liquor and commercial purposes, except boardinghouses and hotels. They sold for $100 to $200 with an easy payment plan requiring 8% interest.

The price of lumber ranged from $14.00 to $20.00 per thousand feet making it possible to construct a four-room cabin for $200.

EXCURSION CARS used on the Alamogordo and Sacramento Mountain Railroad to Cloudcroft.
(Aultman Collection, El Paso Public Library)

EXCURSION CARS.
(Aultman Collection, El Paso Public Library)

Property owners and company representatives conducted government for the community as needed. By April 1900, according to the Alamogordo Improvement Company, North Cloudcroft had livery stables, stores, butcher shops and all the advantages of a country town.

The first Justice of the Peace and grocer in Cloudcroft was A. D. (Punkin) Wallace. Homesteaders in nearby canyons joined the village residents in appreciating the convenience of his store. Mrs. J. A. Buie remembers when it took her father eighteen days to make a trip from James Canyon to El Paso and back for food and other supplies.

In 1901 the railroad company built the Cloudcroft Lodge facing Chipmunk Avenue at the end of Wren Place. Tents across the street, rented by summer visitors, no longer were necessary. This lodge was a two-story structure 212 feet long and 80 feet wide with a slab log finish and stained shingle roof. The dining and kitchen areas faced Possum Avenue forming an ell extension.

Porches completely surrounded the main structure with balconies for guest rooms on the second floor. Rustic in appearance, yet comfortable in furnishings, it attracted guests desiring respite from desert heat in the healthy and balmy mountain air. On the night of June 13, 1909, fire destroyed the Cloudcroft Lodge.

Frank Powers, who built the Southern Pacific building in El Paso, commenced construction of a new hotel in April, 1910. The location of this building was six blocks south of the old lodge site, on Corona. From this high point the view to the west included the White Sands bordered by the San Andrés Mountains many miles in the distance. The highest golf course in the United States stretched to the east.

Mrs. Grace Rogers and her sister Lou thoroughly enjoyed the months it took to build the new Lodge. Their mother operated a boardinghouse for the carpenters. Grace says that she and Lou finished their household chores as quickly as possible each day in order to go to the building project and visit with the carpenters. She recollects most vividly the time that she climbed a ladder to the tower and swung down on a rope. The thrill of this adventure outlasted the rope burns on her hands.

The Lodge, completely fireproof, opened in 1911 under the management of M. B. Hutchins. Guests arrived, sometimes covered with soot, almost at the door. The train left the depot, traveled about two miles up the line toward Russia and then backed to the 1000-foot boardwalk which led to the Lodge.

For the next twenty years, in this tiny mountain village, summer seasons reflected the prosperity of the country and established a precedent for resort living in the Southwest. Many guests, bringing children and servants, stayed for three months. They dined sumptuously at their assigned tables set with sterling silver and Haviland china while listening to music provided by a string orchestra. Oriental rugs and oil paintings enhanced the luxuriousness of the inviting lobby with its huge fireplace. Spacious guest rooms, most attractively and comfortably furnished, completed this setting of elegance. In 1931 the

Southern Pacific Railroad Company, which had bought the Alamogordo and Sacramento Mountain Railroad along with the El Paso and Southwestern, leased the magnificent hotel to Conrad Hilton for $1.00 a year.

The Southwest Lumber Company, in 1934, bought the Lodge along with all timber and lumber rights of the Railroad Company. Louis Carr, owner, hired Rufus Wallingford to operate the Lodge. Eight years later Bill Prestridge, Fred Bonnell and others bought the Southwest Lumber Company and operated the Lodge for the next three seasons.** In 1946, the Prestridge Lumber Company sold the Lodge, sight unseen, to a man from Houston who sold it to C. C. Card within sixty days at a substantial profit. In 1948, a three-man partnership, Bissell, Clark and Allen, bought it and took turns running it during the summer months. As each owner sold, he removed a few of the fine furnishings.

The late John B. Ritter, knowing nothing about hotel business, bought it, in 1953, from a nostalgic point of view. Eleven months later he was killed. James Carnes operated the Lodge from 1955 to 1957 and Wallingford returned until 1959.

At that time, Buddy Ritter, son of John, took over management until a new man could be found. Challenged by the overwhelming possibilities, he is still there. His wife Margaret is the daughter of Fred Bonnell who oversaw management of the Lodge twenty years ago. This charming couple is rapidly restoring the place to its former elegance. Hospitality is not a slogan but a habit of the many employees. Again excellent paintings hang in the lobby. A staff of experts prepare delicious food which is served pleasantly and efficiently in a dining room having a panoramic view of over 200 miles. Guest rooms, most attractively furnished, have modern baths which replace

** Southwest Lumber Company was not sold to M. R. Prestridge until the end of 1945. The lodge may have been sold earlier (1942) as stated.

CLOUDCROFT DEPOT.
(Aultman Collection, El Paso Public Library)

EXCURSION TRAIN, *Cloudcroft, 1909.*
(Orris Smith Collection)

the complete ironstone bowl and pitcher sets of years past. Golf, on the highest course in America, draws players from early spring until late fall. The Red Dog Room, where cocktails and dancing may be enjoyed, is the favorite meeting place of golfers and their friends. Since the Ritter family bought the Lodge, its doors have never been closed.[6]

At the time the old lodge burned in 1909, Dr. Herbert Stevenson of El Paso was trying to interest people who could help him financially in building a baby sanatorium at Cloudcroft. The railroad company had donated land for the project. Dr. Stevenson realized how important it was for some babies to be taken out of El Paso's heat during the summer. However, he knew that some who needed it most could never receive the benefit of cool mountain air because of the expense. Generous donors made possible the building and furnishing of the Baby Sanatorium, which was fondly nicknamed the Baby San. An endowment fund took care of its tiny charity patients.

Dr. Branch Craige, father of the present Dr. Branch Craige of El Paso, and a Miss Knox were in charge of the Baby San when it first opened June 14, 1911. Amazing were the cases of little ones gaining a chance to live and encouraging was the financial report at the end of the first season. After the second season, Misses Emily Green and Louise Dietrich took over administration and care of the infant patients.[7]

For years Mrs. Charles Given gave of her time and energy to the Baby San. Spending winters in El Paso and summers in Cloudcroft, she promoted this wonderful project tirelessly. No doubt it was inspiration from her that helped organize a teenage orchestra. Managed by her son Herbert, these five musicians played for dances at the Pavilion during the summer of 1930 and donated their earnings to the sanatorium. The group used the name "The Chicago Racketeers" because their violinist, Sidney Libit, a nephew of Mrs. Given, lived in that city. Other members were a girl named Kay (?) who played the piano, two Galbraith boys and Edmund Given who borrowed a complete set of drums from station agent J. O. Frilick.

In the early 1930s, as refrigeration became available to families in El Paso, fewer infants became ill during the summer. Gradually, older underprivileged children received unforgettable weeks at the Baby San. Many benefit affairs in El Paso and Cloudcroft helped furnish the ten dollars apiece necessary for one week in the tall pines. During the 1940s this camp was discontinued.

For several years Mrs. La Vora Norman, conducting an art colony, used the building to house students.[8] In 1964, Mr. and Mrs. Ritter built their home on the site of the Baby Sanatorium.

Health and happiness joined hands in Cloudcroft. At the arrival of the first passenger train, and always at train time thereafter, crowds of gay villagers went to the station. Dances held at the Pavilion on Sunday afternoons, attracted crowds until time for the 6:00 p.m. train. At once the dancers rushed to the depot, located on a narrow ledge just below town. Children arrived boisterously fascinated, at times with two and three chipmunks on strings which they sold for thirty-five cents apiece to the delighted passengers of the excursion trains.

Another enterprising young salesman was James Sewell, now operator of the Western Bar (formerly the photography laboratory and curio shop owned by Jim Alexander), who sold the three El Paso newspapers when he was a boy. At the depot he sold several papers, then he boarded the train and rode free up the spur to the Lodge. Knowing it would be ready, he rushed to the kitchen for his customary piece of pie then down the hall to peddle his papers. He says, "I sold more papers in an afternoon than are sold now all summer."

An example of the generosity and kindness of the villagers manifests itself in the story of Joe Billy. No one knew the last name of this man who had helped with the construction of the

LOCOMOTIVE #104 *with excursion train in Cloudcroft about 1902.*
(Courtesy of Mrs. Alexander, Alamogordo, New Mexico)

CLOUDCROFT LODGE.
(Aultman Collection, El Paso Public Library)

railroad. A tree had fallen on him and for years the company took him to Alamogordo for treatment. In spite of his injury and smallness of stature, he was afraid of no one. Rumor even purported him to be Billy the Kid. For years he sold wood which he hauled with a team and wagon. He and his horses lived in unused barns and sheds at various locations in town. Finally the people built him a cabin but did not let him think he was an object of charity. The Myers Company instigated a plan by which he was asked to live in and care for the building until a shipment of special grain arrived from Europe. They are still waiting for the grain.

Once an artist, thoroughly fascinated by Joe Billy's bushy crop of whiskers, made an appointment to do his portrait. Highly flattered and anxious to please, he arrived at the studio wearing a smile but completely clean-shaven. It was the first time people in Cloudcroft had ever really seen his face.

These same citizens were extremely proud and profoundly moved by a concert given in the Pavilion during the summer of 1915. Elizabeth Garrett, daughter of the famous sheriff, gave a recital and for one number she introduced her own composition, "O, Fair New Mexico." The following year the state of New Mexico adopted it as the official state song.

Miss Garrett spent considerable time in Cloudcroft and once impressed Mrs. George Ferguson of El Paso by her philosophy. Mrs. Ferguson, who, in 1900, had been a passenger on one of the first excursions, was most enthusiastically describing some of the surrounding scenery. Miss Garrett, who was blind, said, "Thank you. Some people, when I ask how things look, say, 'Oh, there is just a pile of rocks that make it hard to walk' while others exclaim, 'Such beautiful mountains covered with flowers and trees'!"

Cooperative efforts prevailed throughout the mountain village. All citizens were extremely aware that the signal for a fire was three shots from a gun. Colossal disasters were the burning of the Lodge, in 1909, and the two fires of the Pavilion, the last in 1922. Phoenix-like, they rose again bringing with them

BURNING OF THE CLOUDCROFT LODGE, 1909.
(Courtesy of Emily Kalled Lovell. Photo by Jim Alexander.
Credit: Late C. E. Thomas and Jack Voydé.)

improvements for the entire community. The sewage system, installed the year the new Lodge was opened, was modern in all respects. Of great concern in the second rebuilding of the Pavilion was adequate fire protection. Specifications required either a brick or concrete floor under the range and the installation of two fire hydrants. Completed in 1923, by R. A. Ramey, the Pavilion then had an added dance hall and bowling alley. During the next few months, the company spent $1452.90 installing eighty-five water meters on customers' service connections and over $2000 repairing electric poles.[9] Electricity, from a plant owned by the company, was turned off at midnight throughout the village. Blinking of the lights at fifteen minutes before twelve warned people of approaching total darkness.

Trains almost circled the electric plant going around the U-curve and over the trestle when entering Cloudcroft. Between the trestle and the depot, the spur of Breece Lumber Company connected to the main line. Running parallel to the spur, the highway crossed the track at the east end of the trestle. At one time men were loading a log car on the spur. When they had it almost loaded, they pushed it up to the water tank south of the depot and left it standing alone with brakes supposedly set.** Before long it started rolling, then rapidly gaining speed, it hit the switch and followed the main line right around the light plant. After hitting the derailer, which kept it from going on to Alamogordo, it turned over just ahead of a car coming up the highway. When relating this story, Mr. and Mrs. Frilick agreed that the incident really caused a commotion.

** If the log car was loaded on the Breece spur, the men would not have been able to push it uphill to the water tank. This move would have required horses or a locomotive due to the weight of the car and the steepness of the grade.

DINING HALL AND DANCING PAVILION, Cloudcroft, New Mexico, 1900.
(C. W. "Buddy" Ritter Collection)

There were few station agents and telegraphers at Cloudcroft during the life of the railroad. Among them were Conrad Barrett, Russell Hiller and Joe Frilick. Since there were only two telephones in Cloudcroft for many years, one at the Lodge and the other at the Myers Company, these well-informed telegraphers, like small town telephone operators, were indispensable to the community. Andrew Hendrix, now of Alamogordo, delivered telegrams for Barrett on horseback. Mr. and Mrs. Frilick, still living in Cloudcroft, are authorities on the Alamogordo and Sacramento Mountain Railroad and its mountain village.

Today, children and grandchildren of early cabin owners return to Cloudcroft for the summer months. No gaudy thoroughfare or carnival atmosphere greets and beckons them. Horseback riders mix with tourist traffic on the main street and all are courteous and friendly. Hospitality extends even from the observatory on Sac Peak, nearby, on the steep western slopes of the mountains. Here, scientists, studying the sun, offer guided tours during the weekends for visitors. Graciousness practiced for half a century remains, although the railroad is gone from its trail into the clouds.

CHAPTER FOUR

ROLLING ALONG

THE TWISTING TRAIL OF STEEL reached Toboggan in the spring of 1899. Dedication Day arrived complete with brass band and two locomotives. The Baldwin Locomotive Company had built engine #101, in 1898, for the Alamogordo and Sacramento Mountain Railway Company. This locomotive had two pilot (pony) truck wheels, eight drive wheels and two trailing wheels. A tank-locomotive, it carried fuel and water on the locomotive rather than on a separate tender. Its drive wheels were forty-six inches in diameter. Two cylinders had diameters of twenty-one inches and twenty-four-inch strokes. The total weight of this locomotive was 135,000 pounds. Its working steam pressure was 160 pounds per square inch and its pulling power was 31,290 pounds. (A complete roster of all locomotives used can be found in the Appendix.)

Five years before building engine #101 Baldwin had manufactured engine #102 for their exhibit at the Chicago World's Fair, then they sold it to the Alamogordo and Sacramento Mountain Railway also. It had only four drive wheels and was much smaller than #101, weighing but 72,130 pounds.

For these two trains which went to Toboggan twice daily we have the following schedule from the *White Oaks Eagle*[1] of January 4, 1899.

El Paso and Northeastern and
Alamogordo and Sacramento Mountain Railways
Time Table No. 2

Train No. 1 leaves El Paso .... 10:30 A.M.
Train No. 2 arrives El Paso .... 6:50 P.M.
(Daily Except Sunday)

Trains leaving El Paso on Mondays, Wednesdays and Fridays make through connections to Capitan.

Trains arriving at El Paso Tuesdays, Thursdays and Saturdays have a through connection from Capitan leaving there at 10:30 A.M. and Carrizozo at 11:00 A.M.

Trains Nos. 1 and 2 run via Jarilla, the great gold and copper camp on Tuesdays and Fridays.

Trains leave Alamogordo for Toboggan on the summit of the mountains, twice a day.

STAGE CONNECTIONS

At Tularosa—For Mescalero Indian Agency and the San Andres mining region.

At Carrizozo—For White Oaks, Jicarilla, Gallinas and the surrounding country.

At Walnut—For Nogal.

At Capitan—For Fort Stanton Sanitarium, Gray, Lincoln, Richardson, Ruidoso and Bonito country.

At Toboggan—For Pine Springs, Elk, Weed, Upper Penasco, Penasco, and the entire Sacramento Mountain Country.

For information of any kind regarding the railroads or the country adjacent thereto, call on or write to

A. J. Greig
Gen. Supt. & Gen. F.&P. Agt.
Alamogordo, New Mexico

or H. Alexander
Asst. G. F.&P. Agt.
Alamogordo, New Mexico

or F. E. Morris
Local Agent
El Paso, Texas.

ENGINE *#101–102 at Pinto Dedication Day, 1899.*
(Museum of New Mexico, Santa Fe, New Mexico)

In 1899, Baldwin built two more locomotives for this railroad, engines #103 and #104, each weighing 140,600 pounds. They were originally ordered as 2-8-OT (tank locomotives), but were built with tenders. Pictures showing huge balloon smokestacks and wood racks prove that they all used wood for fuel. The name Alamogordo and Sacramento Mountain Railroad Company stretched proudly on the side of each.

Because of the extreme grades and curves of the track, the use of geared engines was seriously considered during the first few years of the railroad's operation. Shays of the Alamogordo Lumber Company regularly supplemented the railroad company's standard rod-type engines. This experimentation with Shay-type engines finally culminated in the purchase of the first 4-truck Shay #105 which Lima Locomotive Company built in 1902, that was used the next year on the mountain railroad.

This locomotive weighed 260,300 pounds and was the world's largest at the time.

A Shay engine was a geared steam engine that had a bank of three cylinders on the right side of the boiler that drove a crankshaft similar to one in an automobile engine. From there power was transmitted to all wheels by means of driveshafts and a gearbox on every axle. The gears were arranged so that the engine was always in low gear. Top speed for a Shay engine was fifteen miles per hour. Although it was slow, a Shay had tremendous pulling power and could pull heavy trains up grades no conventional engine could conquer. Because of the weight of the cylinder bank and driveshafts on the right side, the boiler was slightly off-center to the left to keep the locomotive balanced. Thus, it presented a rather unusual appearance

ENGINE #102 at Toboggan on Dedication Day—a wet celebration.
(Museum of New Mexico, Santa Fe, New Mexico)

when coming head-on. Sold in 1905, Shay #105 became Ferro-carril Mexicano #110 (National Railway of Mexico #110).

Apparently the use of Shays was deemed unnecessary and for several years rod engines, requiring less maintenance and care, were used exclusively. However, one last trial using a Shay was made ten years after #105 had been sold. This 4-truck Shay, #99, built in 1907, became property of the El Paso and Southwestern Railroad (successor to the El Paso and Northeastern) in 1915. It was larger than Shay #105, weighing 357,300 pounds.[2]

Shay #99 made only one round-trip to Russia. Engineer Jim Riddle and Conductor John Tweed were unable to complete the trip in sixteen hours and had to stay at the switchback overnight. On the way up the mountain Riddle had burned his arm. At the switchback he went to sleep in the coach while Tweed made out the report of his injury. The next day Riddle, still infuriated, grumbled to Agent Ernest P. Rees, "Do you know what that old so-and-so did? He woke me at three in the morning to find out how old I was."

The engine was too long for the great curvature of the track. It was sold to the Red River Lumber Company of California in 1920.

A series of four Baldwin locomotives supplemented earlier ones. These, built in 1900, for the El Paso and Northeastern, weighed 141,000 pounds. Numbers changed as the railroad changed ownership. Numbered for the El Paso and Northeastern as #52 to #55, they became #181 to #184 for El Paso and Southwestern. A picture from the collection of Orris Smith of Capitan shows engine #184 on its back off the track near Wooten. Engineer F. M. Weldey was killed in this accident.

Two other Baldwin locomotives, numbers 19 and 24, built in 1902 and 1905, finally became Southern Pacific #2510 and #2511. These engines weighed 176,000 pounds each and like all Baldwins except engines #101 and #102 had no trailing wheels. Engines #2507, #2510 and #2511, modified with special couplings,

ENGINE *#101—Alamogordo and Sacramento Mountain Railroad Co. BLW 16103 1898 2-8-2T 46-21x24-135,000-160-31290. Scrapped 7-24-34.*
(Vernon J. Glover from original B. L. W. negative, H. L. Broadbelt Collection, Hershey, Pennsylvania)

air brake equipment and wide driver tires, were assigned permanently to the mountain railroad.

S. A. Ramsdale of El Paso says engine #2510 was the meanest engine he ever ran. It once barely missed going into a large hole where water had run under the track. Ramsdale's own words tell exciting experiences with that engine.

> "We were going along on a straight track and for some reason the rail on the left-hand side turned over and the engine jumped the track. She wobbled and jumped around and, when we finally got her stopped, she was back on the rail. But there had been two wheels, one on each side, in between the rails, which had turned over the left rail. From the tank back eleven car lengths there wasn't a wheel on the ground. Every left wheel was on the web of the rail. We spiked a frog between the engine and tank and pulled the whole train along with the wheels on the flange of the rail. Pulled them up to the frog and they all went back on the track. We went right on to Russia, then we called a section gang. They straightened the rail out and spiked it to the ties while we were gone. We came back over that track and went on to town."

Working close to Grand View Curve, Wiley Smith, section foreman, watched #2510 as it went off the track. He arrived as the second whistle finished blowing, about ten minutes after the accident. He took the mail to Cloudcroft on a motor car. By the time he returned, the trainmen, using frogs, had the engine rerailed. It was then his duty to repair the damage done to the track.

A rerailing frog is a steel casting shaped to guide the flange of the derailed wheels back to the correct position on the rail. It is spiked in place under the derailed car and the car is then pulled slowly along. The derailed wheels catch on the frog and

ENGINE *#102—Alamogordo and Sacramento Mountain Railroad Co. 2-4-2T #13361 4-1893-44-14x24 72130 Baldwin Locomotive Works.*
(Photo by Jim Alexander. Courtesy of Mrs. Alexander)

roll over it. Thus they are raised up onto the rail and dropped down onto it in the proper position.

Sometimes trains would derail in the middle of the high bridges. It was then necessary for the trainmen to crawl along the trestle and set frogs to get the cars back on the track.

> "In 1923, I was firing," Ramsdale continues, "and we had a big snow storm. We left Alamogordo at 7:30 that morning and got about a mile out of town when the wind blew the right glass out of the window on the engineer's side. The engineer had to stand in the snow and the storm all the way to Wooten. Got there and got an apple box and nailed it over the window to keep the snow out. Got to Toboggan and picked up the plow car. When we left the top of the switch-back we ran into a tree that had been blown down across the track. We had picked up all the section gang, two groups with us. They got out and sawed the tree in two pieces, got a chain around it and pulled it off the track. We got on our way and just before we got to Dark Canyon, close to the "S" bridge, below Cloudcroft about a mile there were four big trees lying across the track. The men got out, sawed them in two, pulled them off the track and we got on our way again.
>
> "We got to Cloudcroft, went on up below the Lodge to the Wye to turn around because we could not go any farther. Had to get back—sixteen hours. There was a small tree across the track so we sawed that in two, got back to Cloudcroft and it was 6:00 p.m. after leaving Alamogordo at 7:30 that morning.
>
> "Went up and ate. Engineer went to caboose and sent brakeman to fire for me down the hill. When I turned the headlight on, I had no light, so I came all the way down the hill with the caboose with no headlight. At that same place where we had pulled off four trees going up, we had to stop and pull off five more that had fallen from above the track up the hill and their roots with rocks frozen to them were in the middle of the track. I hit that with the engine because I could not see it in time to stop, but I did not do any damage.

"My last experience with this engine #2510 was during the winter time when we had been running on snow and ice all the time. Just west of Cloudcroft going along one morning I just had a caboose. Right on the curve there was a crossing and trucks and cars going across it had covered it with snow and ice. When we hit that left-hand curve, making about ten miles per hour, instead of making the curve, the engine left the track and ran off into the side of the hill and almost turned over. I jumped off in snow almost waist deep. She was leaning to the right. I was afraid to get back on it to shut the lubricator off for fear my weight would make it turn over. I hollered at the fireman, when she left the rail, to jump, but he didn't understand what I said, so he rode on her and, fortunately, she did not turn over. We had to put the fire out, blow all the water out of the boiler, open all the valves, drain everything and let her stand there. We went back by truck to Alamogordo. Next morning came back with another engine to re-rail her. They had to cut the rail and build a railroad under her to pull her back straight because she was so far from the rail."

During winter months a snowplow was part of the necessary equipment. A wedge plow, fastened to the front of the engine or to a log car loaded with scrap iron, threw snow to each side of the track. Jake Jacobson built the four snowplows.

Station agent J. O. Frilick says, "One time we lost a boxcar right here in Cloudcroft. It was completely covered with snow on the side track. I kept getting tracers on that boxcar. The last record they had of it was up here, you know. Of course, the conductor should have left a blind siding report for me. Lots of times conductors didn't do what they were supposed to do. So the last record was that it left Alamogordo to come to Cloudcroft. I had no record of it, so we just lost that boxcar. When spring came and the snow melted, it uncovered the boxcar."

One time during the winter, when H. S. Fairbank was train master, he decided to show Jim Riddle how to get through a snowbank on a curve. Riddle felt it was impossible to make the

ENGINE #99—*El Paso and Southwestern's last 4-truck Shay made only one round-trip to Cloudcroft.*
(O. Rasmussen Collection)

ENGINE #2505-181—*Taken below High Rolls. Only place where S. A. Ramsdale could get picture of entire train in mountains.*
(S. A. Ramsdale Collection)

curve, but Fairbank said, "We'll just ram her into it." When he suited action to his words, the locomotive went straight ahead and turned over in the snowbank.

Livestock as well as snow caused hazardous conditions on the track. The engineers frightened cows and other animals off the track by use of steam and the accompanying roaring noise made by blowoff cocks, one on each side of the firebox, controlled by a lever in the cab. The real purpose of these cocks was to clean sludge from the leg of the boiler.

Engineer Jim Riddle invented the rail washer, used to keep the tracks clean. A pipe connected to the boiler carried water in front of the pony trucks. When valves were opened, this hot water wet the track. A stream of sand was then blown onto the track. If necessary, a second stream of water behind the engine washed off the sand, lubricating the rail so the cars would not derail. When the company would not buy his invention, Riddle removed it. Proof of its value appeared daily and soon the company bought the patent for $1500. By that time, another engineer, F. M. DeLap, had perfected the same invention. Riddle received the franchise for the western part of the country only, while the other inventor reaped benefits east of the Mississippi River and the honor of having the invention named the DeLap Rail Washer.

During heavy rains sand from the sand dome provided the necessary friction to keep the wheels of the engine from spinning. Mrs. A. T. Seymour says that the engineer must have been out of sand on one trip when her mother, Mrs. W. A. Hawkins, was on the train. Conductor Tweed went through the coach asking for any dry supplies that the passengers might have. Mrs. Hawkins had a large cannister of tea which, when poured onto the track, worked perfectly.

Mrs. Mary Dozier of El Paso, widow of Fireman R. W. Dozier, remembers living in Cloudcroft during the summer when rains washed out a bridge and almost wrecked the train

ENGINE #2506—Alamogordo and Sacramento Mountain Railroad Co. on S-bridge, 1936. Water barrel on right of bridge. (E. Clack Collection)

on which her husband was firing. She has never forgotten the excitement of that trip of which the *Otero County News*[3] of August 7, 1914, gave the following account:

CLOUDCROFT TRAIN HAS NARROW ESCAPE

Heavy rains in the Sacramentos during the last thirty days have done much damage. The one which fell Wednesday afternoon washed mud sills from the bridge about three miles above La Luz on the Alamogordo and Sacramento Mountain Railroad thereby almost paving the way for a serious accident. No. 22, with about 45 passengers, was unable to cross the bridge Wednesday evening and the train returned to Alamogordo where the passengers spent the night. Just as the pony trucks of the engine passed on the bridge the absence of

> the mud sills made the bridge settle several inches. Engineer William Wade McLean felt the bridge settling but did not dare stop the engine on the bridge. He kept moving and crossed without any further trouble other than the derailment of the engine, just as it reached the further side. The bridge was temporarily cribbed up so that the engine could get back across and the return trip to Alamogordo was made, the train arriving here about 11 P.M. A work train started early Thursday A.M. to make necessary repairs and the passenger train started again for Cloudcroft at 9:15 Thursday A.M.

Mrs. Dozier remembers that the train, controlled by brakes, was backed down without the engine. Passengers included Mr. and Mrs. Van Swearengin on their honeymoon.

The Alamogordo and Sacramento Mountain Railroad might well have been called the Honeymoon Special. Among the many couples who experienced pleasure from a wedding trip on it were Mr. and Mrs. Tom Bell. Some years later the Bells were living in Mountain Park when Mrs. Bell's sister with five children came to visit them. With baby in arms, plus children, luggage and other paraphernalia, she alighted at the station. Suddenly she realized that the train had started on with her two-year-old daughter still in the seat asleep. Bell jumped into his pickup, raced down the highway which was parallel to the track and passed the train. He leaped on it as the train slowed for a curve, grabbed the child and jumped off. Neither Engineer Riddle nor Conductor Tweed thought it necessary to stop the train. Twelve miles an hour was the speed limit.

Tom Shorten, retired and living in Weatherford, Texas, seemed to have expressed the general opinion of all engineers concerning speed on the line. Someone once asked, "Tom, how do you like the mountain railroad?"

"Well, if I go too slow the logs jump off and if I go too fast the cars jump off."

ENGINE #2507 and crew, 1936.
(E. Clack Collection)

AT BAILEY'S with stock car, 1936.
(E. Clack Collection)

C. V. Long, for two years a fireman, then some time later an engineer on the mountain route, asserts that the trip from Russia to Alamogordo could be made in three hours. Often it took sixteen to eighteen because the log cars rocked off the track. In his estimation the job of brakeman was the most hazardous because, when the stay chain broke, the logs rolled off and hit the man beside the track.

W. L. (Billy) Smith, father of Orris, began working as brakeman in 1901. He had one hand crushed when coupling two cars. In 1923, as conductor, he was helping the brakeman give signals in a snowstorm at Russia. A log rolled off a car, pinning him beneath it and breaking his back. Jesse Bays, assistant section foreman, and one other man lifted the log while E. P. McCrary helped remove the injured man. When officials checked the accident the following day, seven men could not lift that log. Smith spent time in the hospital in California and wore a steel brace for the rest of his life. For three years before his retirement he supervised the railroad clubhouse in Carrizozo.

The brakeman's job, although dangerous, was indispensable. Mrs. Frilick relates the following story of a night during their first week in Cloudcroft:

> "The train crew had left the train between the depot and the water tank and had supposedly set all the hand brakes, but evidently they had not. Well, they came over to pick up the logs right down here from main street, right here in Cloudcroft. Evidently they didn't set more than every other one or didn't set them good and the darned thing turned loose. They all ran down there. We heard this commotion and first thing we knew we heard this terrible crash. Log cars turned over and piled up onto the platform and broke it down. My husband and another man ran over there. We'd been sleeping on pallets on the floor waiting for our furniture to arrive. We had to sit up all that night with the train crew because they had nowhere else to go and it was snowing so bad.

> Could not get out anyway. The next day they brought up an engine from Alamogordo and pulled all those logs off the platform, and those cars—got them back on the track. They had to leave the logs a few days until they could get the loader up here. It was kind of a scary experience."

Brakes were as important as fuel, Sy Ramsdale says:

> "Coming down the switchback with a twenty-two-car log train, the track at the bottom was about six per cent grade and you had to come off the switchback as fast as you could stay on the track in order to shove the train up this heavy grade. In going down you got at a certain place at the bottom that you needed to release all brakes so there would not be any dragging and the conductor would open the retainer line on the caboose and that would bleed out all the air. Then you'd start working steam and as soon as the air quit blowing out he'd close that valve. Then you had to set your air immediately or the train would run off with you."

The retainer line, used only on cars in the mountains for additional control, could be released by a valve in the engine or from the caboose. A soap and water test was the most satisfactory way of detecting leaks in this line. When a brush full of soapsuds was used on its connections, bubbles would show leaks that could not be heard.

Going up a hill it was never necessary to set a brake to stop. The engineer shut off the engine and the train would stop itself. It was necessary to set brakes on the engine to keep it from rolling backward. When he had made a stop going down hill, to start again, the engineer simply released the brakes.

There is at least one instance of a caboose, controlled by brakes, traveling for some distance without an engine. J. P. Nash, train master, was on an excursion train leaving Cloudcroft after a Fourth of July celebration. There were several cabooses on the train helping carry passengers when one developed trouble at El Valle.

Loop at Bailey's, 1936. Engine #2506 is cut into middle of train.
(E. Clack Collection)

Excursion Train between La Luz and High Rolls about 1915.
(John M. White Collection)

Engine #2508-184 *lying wrecked above Wooten Tank about 1910.*
(Orris Smith Collection)

Engine #2510—*Lucky accident—right front wheel pony truck axle broke.*
(S. A. Ramsdale Collection)

Nash finally sent the rest of the train down the hill. He, with the assistance of one brakeman, took the caboose to Alamogordo.

Adequate braking equipment on cars was also extremely essential. After finishing the railroad to Cloudcroft, the company needed ballast for the track north to Tucumcari. Mexican laborers were filling a gravel car which had been placed in La Luz. Instead of carrying the gravel to the car, the men decided to roll the car to the gravel. They removed the rock which was holding a wheel and released the obviously defective brakes. Slowly the car rolled to the pile of gravel and on beyond, gradually picking up speed. Realizing that it was impossible to stop the car, the men on it jumped off at the first curve. On it went toward Alamogordo, just as the passenger train was arriving from the south. Seeing the impending danger, someone threw the switch and the car side-tracked just west of the depot. Here it ran into a string of cars, which was waiting for another train, and wrecked. Fainting relatives of passengers on the train, expecting them to be injured or killed, had to be revived on the station platform.[4]

According to the agreement signed December 6, 1898, between the Alamogordo and Sacramento Mountain Railway and the Alamogordo Lumber Company, the lumber company was to provide suitable and a sufficient number of log cars. On these a straight air brake supplemented the usual automatic brake. This made possible a continued brake pressure on the grades downhill. The log cars had to have braking appliances that met requirements of the railroad.

The railroad company also determined the number of cars an engine could pull. Unless a second engine or helper was used, engines #2510 and #2511 pulled no more than fifteen cars, while the other engines could pull only eleven. At the switchback, when there were over twenty-two cars, the train had to be cut in two. Usually there were eleven cars between the two engines and ten between the second engine and the caboose. Going up, the whole train would go to the lower end of the

ENGINE #2511—hauling logs.
(E. Clack Collection)

switchback. After the front cars were cut from the second engine, the first engine would pull up, back its eleven cars to the top of the switchback, then head out toward Cloudcroft. When it got out of the way, the other engine, cars and caboose would pull up, back up the switchback, go up behind the first section of the train, couple up and the whole train would go to Cloudcroft. Coming down into the switchback it was necessary to stop not over three feet from the mountain in order to throw the switch. If the engineer did not estimate correctly, the engine went into the hillside. Ernest Rees, retired agent at Alamogordo, says, "Many a pilot (cowcatcher) was left at that spot."

Ramsdale had another experience best told in his own words:

> "One time I went up with a little engine, one car of coal and a caboose. About 300 yards before we got to the bottom of the switchback was a 33-degree curve and about 4 per cent

grade. I couldn't pull up one car of coal and one caboose around that curve. I had to cut the caboose off and we left it there. We took this car of coal up to the bottom of the switch-back, set it on the side track, went back and got the caboose and pulled up there and set it on the main line. Then we went back, picked up the car of coal, backed up against the caboose and went on to Cloudcroft."

Making up a train at Cloudcroft was another complicated procedure. Having left the empty log cars, the engine and caboose came down from Russia to the Cloudcroft depot, where the caboose remained until loaded cars were coupled to the locomotive. Just below the depot was a wye having one straight leg and one which entered the curve just below town. A tail (back) track held boxcars loaded with lumber from local sawmills. The engine went beyond the track which connected the legs of the wye, backed onto it and up the straight leg of the wye. There stood the cars loaded with logs which the engine pulled onto the track between the legs of the wye. The brake-man then moved the boxcars, loaded with lumber, one at a time, off the back track, to the rear of the train of log cars. When they were all connected, the engine pulled the whole train, except the caboose, onto the main line. Last, the conductor coupled the caboose to the train.[5]

Their many deeds of kindness endeared these trainmen to the children living near the mountain railroad. Mrs. A. P. Sitton, formerly Josie Ritchie, of Nogal lived in Cloudcroft as a child. She was about seven when she learned that she could earn her fare from below the "S" bridge to the town. During the summer there were many wild strawberries growing near the track. After eating her fill, she needed only three berries for Conductor John Tweed to allow her to ride home.

As a boy about twelve Raymond Buckner had gone to visit the Bonnells at Russia. He had spent all of his money for candy except the amount he thought necessary for return fare to Alamogordo. When he boarded the train, the conductor told

ENGINE #103 Alamogordo and Sacramento Mountain Railroad Co. with tender filled with wood, 1899.
(Vernon J. Glover Collection)

ENGINE #103 "Modified" Alamogordo and Sacramento Mountain Railroad Co. EP & NE 103-EP & SW 185-CL & LCo. 1 on April 24, 1924, to George E. Breece Lumber Co. #1 in 1926.
(S. A. Ramsdale Collection)

him that he had only enough money to ride to La Luz. Although worried over the prospect of a five-mile walk at night, lulled by the swaying of the train, the exhausted boy fell asleep in the corner of the caboose. When the train stopped, he awoke, scrambled to his feet and grabbed his little satchel. Fearfully he stepped off the train into the darkness, thinking he was at La Luz. The train was at the depot in Alamogordo.

Children in Cloudcroft often worried the trainmen by hanging on the back and sides of the tenders. Even so, the men were kind to them. Agent Ernest Rees once answered, "I don't know," when asked about Engineer Kelly's having two children on an engine with him.

It was little wonder, though, that the section man hid the third wheel of his three-wheel handcar from the village youngsters. However, one morning Edmund Given and Junior Lusk found and attached it. The next step was inevitable. On they climbed and down the track they sped. Too frightened to jump, neither dared try to grab the handlebar that was pumping madly back and forth. There was nothing to stop their wild ride until they reached the switchback.[6]

What a pleasure the five children of Conductor E. P. McCrary experienced when he permitted them to ride in the caboose! So exciting was the ride that the youngsters usually stood during the trip. His daughter, Mrs. Wilbur Fifer, says that the sudden jerks and swaying of the train caused frequent bumps on their heads. One visiting cousin hit the door of the caboose with such force that his head went through a panel. The greatest thrill resulted from being allowed to sit up in the cupola, which would hold four children, two on each side. Each caboose had long seats on the sides that would seat eight or ten men; a desk for the conductor with kerosene lamp fastened on the wall nearby and a potbellied stove, which furnished heat, standing in the corner.

Accommodating acts of trainmen were not limited to the children along the route. Errands for the entire family could range from purchasing a bottle of cough syrup for a baby with

ENGINE #105 Alamogordo and Sacramento Mountain Railroad Co. Shay with 42 cars. "S" between El Valle and Pinto. (Courtesy of Vernon J. Glover, Jr., Lima Collection, P. E. Percy)

the croup to selecting a suit of woolen underwear for an aging grandfather. The grateful recipients or their relatives expressed their appreciation with gifts of fruit, vegetables and venison. Engines of Jim Riddle and Bob Dozier often left High Rolls completely bedecked with flowers.

The Railroad Company had one car which ran daily except Sunday, a combination passenger-baggage-express-mail and caboose. During the week this car and an excursion car would go to Cloudcroft and back up to the Lodge. There, these two cars remained overnight. In the morning, the engine would go to Russia, take the log train back to Cloudcroft, cut it off and leave it there, back up to the Lodge and pick up the passenger cars and go to Alamogordo to meet the westbound passenger train. Women from El Paso who with their children spent summers in Cloudcroft used to sit at the station and wait for the late afternoon train on Fridays. On it their husbands would arrive from the city for the weekend.

On weekends and holidays, during summers a train of ten or twelve passenger cars from El Paso arrived in Alamogordo. Here, its passengers, joined by others from Alamogordo, boarded one of the four yellow excursion cars in order to spend a day or more in Cloudcroft. An excursion car, built in Alamogordo and called a Balley-Claire, was about the size of a boxcar and open on each side. It had curtains that could be lowered in case of a storm, but most of the sightseers, enthralled by the magnificence of the scenery, disregarded a few drops of rain. Each car could comfortably hold forty-five passengers but on holidays such as July Fourth, when a regular passenger car was often added, as many as six hundred passengers squeezed into and on top of the train. Flatcars, equipped temporarily with benches, and even cabooses sometimes held the overflow.

The excursion trains left Alamogordo about nine o'clock in the morning. Gaiety prevailed throughout the crowd which was decked in summer finery. Many of the passengers carried baskets filled with picnic lunches while others planned to eat at the Lodge or the Pavilion. At each stop along the way a crowd of spectators greeted them.

Too excited to be frightened, most of the inexperienced travelers joined the sightseeing and merrymaking of some who had been the railroad's earliest passengers. From the middle of a swaying trestle, looking to the top of a towering escarpment or glancing at the floor of the canyon below, no doubt, a few wondered if they would ever live to tell of the spectacle and, if so, why. Even the depot at Cloudcroft seemed to hang on a narrow precipitous ledge.

At that station, agog with anticipation, thronged the entire population of the town. Fashionably dressed men and women, children on burros, babes in arms and even the dogs greeted the visitors.

The railroad village had facilities for golf, bowling, horseback riding and dancing. Some of the guests preferred the luxury of doing nothing but breathing the pure mountain air or viewing the unspoiled beauty of the glorious country.

An advertisement in the *El Paso Times*[7] on August 31, 1907, offered a round-trip for Sunday or Monday at $3.00. The Fort Bliss 19th Infantry Band was to accompany the excursion, remain over Labor Day and give a series of concerts. The height of the entertainment was to be a Masquerade Ball to be held Monday night. Passengers could go and return the same day or remain until Tuesday. Since it was the custom for the entire community at any of the stations along the line to meet the trains, excitement must have known no bounds on that particular holiday weekend.

Among passengers of one excursion for a day in Cloudcroft were Jenny Garvin, her brother, his lady friend and William McLean. Jenny, now Mrs. Dolan of Tucumcari, remembers wearing a white linen dress while the other girl wore one of brown. McLean had a Kodak and took a picture of the girls leaning against a stump. They had not noticed the pitch oozing from it until their dresses were stuck to newly cut pine and the white one badly stained. It was on this trip that William proposed to Jenny. She says, "We spent much of that day trying to get away from my brother."[8]

From the depot platform that was again overflowing, the train departed at 6 P.M. for Alamogordo. There the El Paso-bound passengers boarded the regular train. When it arrived in the northeast part of the city, it made several stops to accommodate those living in that area.

When Charles B. Eddy** sold the Alamogordo and Sacramento Mountain Railroad, along with his El Paso and Northeastern, to Phelps Dodge in 1905, his mountain railroad listed

---

** Eddy, who never married, continued his career as promoter for the next twenty years, on this continent as well as in Europe. His enterprises included underground telephone conduits in Chicago, mining activities in Mexico, railroad construction in Spain and oil ventures in Texas. He died in New York City on April 13, 1931.

El Paso and Southwestern System

ALAMOGORDO AND RUSSIA

Leave (Mountain Time)				Arrive
No. 22	Miles	September 17, 1922	Alt.	No. 21
700 AM	0.0	Alamogordo	4320	250 PM
725 AM	6.0	La Luz	4836	225 PM
745 AM	9.7	El Valle	5380	210 PM
800 AM	12.6	Pinto	5953	155 PM
815 AM	15.8	High Rolls	6550	135 PM
905 AM	20.6	Toboggan	7728	100 PM
950 AM	26.2	Cloudcroft	8600	1225 PM
1005 AM	28.1	Cox Canyon	8500	1125 AM
1030 AM	32.5	Arr. Russia	Lve	1100 AM

g. Monday, Wednesday and Friday

the following equipment: four locomotives, one combination - car, four excursion cars and five cabooses.

The railroad then became part of the El Paso and Southwestern System. The following schedule appeared in the Official Guide, May, 1923.

The United States Mail contract required that this trip be made three days a week.

A year after purchasing the El Paso and Southwestern Railway from Phelps Dodge, Southern Pacific was using six locomotives on the mountain line and, during the summer months of 1928, scheduled three trains a day except on Sunday.

For years, a track rider followed the trains. He rode horseback through the forest to extinguish any fires that might have started from sparks or ashes of wood and coal. The horse could cross the bridges which were floored and covered with gravel. Until locomotives started burning oil as fuel, the caboose was not the end of the train.

CHAPTER FIVE

Logging

FROM TIME OF ORGANIZATION to removal the very existence of the cloud-climbing railroad depended on logging, first for ties for the main line, then as a profitable industry. Inhabitants of Lincoln County had been vaguely aware, for years, of a vast expanse of timber in the Sacramento Mountains. In addition to private owners, the Mescalero Indian tribe, the National Forest, and the Territory of New Mexico all had timber holdings in the area.

Less than two months after construction began on the Alamogordo and Sacramento Mountain Railroad, the Alamogordo Lumber Company, on May 19, 1898, obtained a charter with $200,000 capital stock.[1] The lumber company built two mills, one late in 1898 and another the following year, each equipped with the latest and most improved machinery. The railroad company and the lumber company put into operation every appliance available in the late nineteenth century for cutting and hauling timber to the mills in Alamogordo.

By December 6, 1898, C. D. Simpson, as president of the Alamogordo and Sacramento Mountain Railway, entered into an agreement[2] with F. L. Peck, president of the Alamogordo Lumber Company. Terms of this twenty-five-year agreement are here greatly condensed:

I. Lumber Company would build and equip sawmill capable of sawing 50,000 board feet per day (11 hours).

II. Lumber Company was to supply energy, capital and labor to keep sawmill running.

III. Lumber Company would deliver all timber accessible to railroad lines which went to mill in Alamogordo.

IV. Lumber Company would build necessary laterals and tramways to deliver maximum amount of logs to railroad.

V. Lumber Company would provide suitable and sufficient number of cars (110).

VI. Railroad Company would have right to build line over Lumber Company land.

VII. Railroad Company would construct and provide a railroad from mill to summit of Sacramento Mountains and would extend it, if necessary, for supplies of timber. Would not build extensions over any route of over 5% maximum grade; or which required expensive bridges, fills, cuts, tunnels and other structures.

VIII. Lumber Company had right to require from Railroad Company necessary connections between its laterals and tramways and the railroad lines. All switches were to be under Railroad supervision and expense.

IX. Railroad agreed to receive all logs delivered to it and loaded safely, and to transport to mill at Alamogordo. Railroad would return empty cars to points designated by Lumber Company. Lumber Company log cars had to have braking appliances as required by Railroad. Railroad was to pay no mileage or rental on cars.

X. Charges:
   1. Logs—$2.00 per 1000 board feet.
   2. Shingles, laths, etc., fair rate according to selling price of merchandise.
   3. Materials and supplies used by Lumber Company in mountains and at mills, $1.00 per ton in carload lots, 24,000 pounds minimum.

XI. Smaller lots for material going rate—50% rebate given.
XII. Statements to be sent 10th day of month.
XIII. Disputes would be arbitrated by chosen board.

This agreement* was signed by Simpson and Benjamin S. Harmon, Railroad Secretary and Peck and T. H. Watkins, Secretary of the Lumber Company.

The earliest logging operations, using teams to haul logs to the railroad, started before completion of the switchback. They penetrated the forest both east and north from Toboggan to the south boundary of the Mescalero Indian Reservation. Moving timber to Toboggan was a strenuous task. As many as six yoke of oxen pulled a single log to Sally White's sawmill or to the railroad. Often four or five logs, tied one behind another, were dragged by the team. Drivers of oxen, known as bullwhackers, were a breed unto themselves. Extremely skilled in the use of the bullwhip, they were able to control the team by barely touching one animal. "Flicking a fly from the tip of a horn" may not have been an exaggeration of the bullwhacker's ability.

A nickname for Toboggan was Slabtown. There, Mr. and Mrs. John E. C. Bell, parents of Tom Bell of El Paso, owned an eating place not far from White's sawmill. A room adjoining the kitchen served as undertaking establishment for the community. It was not unusual for the loggers to settle their differences permanently. Tom Bell remembers the night that Tom Sanders and a logger named Ravencraft argued over fifty cents in the local saloon. Sanders lost the argument and his life.

---

* In files of the Southern Pacific Railroad offices, along with this contract, are two letters. One was dated June 27, 1905, stating that the El Paso and Northeastern Railroad had set the rate of lumber at $4.00 per 1000 feet. A curt reply on June 30, answered that the above contract of $2.00 was for twenty-five years. The next day Phelps Dodge completed transactions for the purchase of the El Paso and Northeastern Railroad which became the El Paso and Southwestern.

LOG LOADER.
(Orris Smith Collection)

G. E. BREECE LUMBER ROAD—Log Loader.
(U.S. Forest Service)

The next morning Tom ate his breakfast with the sheriff and Ravencraft while looking at the corpse on a board in the next room. The killer finally said to Tom's mother, "Well, Mrs. Bell, this will be the last meal I'll ever have with you." The sheriff's prisoner escaped on the way to Alamogordo.

As the main line advanced to Russia, beyond Toboggan other lumber camps began operation along the way. The second camp, just below a hairpin curve, was Bailey, which was named for its first foreman. Two logging railroads joined the main line there and W. L. Rutherford owned its sawmill.

The Pettijohn Camp, also named for its foreman, was in James Canyon where the Cloudcroft Water Works is located. Mrs. Pettijohn, wife of the foreman, shocked the early inhabitants of the canyon. Often, on horseback, she look a lunch to her husband. Instead of using a sidesaddle, she rode astride. The Alamogordo Lumber Company had a logging railroad going into James Canyon. Also, on down James Canyon, on land owned by a Mr. Painter, was Painter Camp.

Logging started in Cox Canyon as soon as the railroad reached Cloudcroft. On October 19, 1903, near Cloudcroft was the Alamogordo and Sacramento Mountain Railroad's most tragic accident. A steel crew was working on final construction near Russia. Their train left the track at the head of Cox Canyon and two cars, loaded with steel, telescoped. Seven Mexican laborers, sitting on top of the steel, lost their lives.

From Cloudcroft to Russia was a distance of six miles. About sundown one afternoon, Carl Lenz, a Polish logger, after having spent considerable time in the bar at Cloudcroft, started walking to Russia carrying his little black hand satchel. Lenz had often talked about ice-skating to the other men. About two miles from the end of the line, exhausted or intoxicated, he lunged off the track into a bank of snow piled there by the snow plow. In it he spent the night with only his feet completely exposed to the cold.

3T SHAY #2, 1902.
(Orris Smith Collection)

LOGGING TRAIN in Lincoln National Forest.
(O. Rasmussen Collection)

The following morning the trainmen, on the trip from Alamogordo, found him and took him to Russia. He realized his feet were frozen and kept repeating, "Poor Carl, poor Carl, never can skate some more."

Later, at the hospital in Alamogordo, where his feet were amputated, someone checked the contents of his bag. It contained three gold medals, attesting him to have been the champion skater of Poland.[3]

Boilermaker Joe Moore and his apprentice, Carroll Woods, made a trip to Russia in 1904 to inspect the lumber company's Shay engines. The boiler of one had exploded and killed Lee McNatt. Moore condemned all the locomotives and tied up all activities until the engines could be reboilered in the shops at Alamogordo. When the two arrived at Russia, there was snow six to eight feet deep in places. From there travel became impossible except over a trail to the west side of the summit and down the mountain.

Mat Massey was the blacksmith who made necessary repairs on the Shays and logging equipment. He was the proud possessor of a new Winchester 35 Special rifle. Two days before Woods and Moore arrived at Russia, he had taken the trail to Alamogordo, then on to Rincon with two horses and the gun, which he jokingly called his "twenty-two." After a trip lasting five days, he returned with horses loaded with deer. His report was, "Boys, this old 'twenty-two' is shore a dandy. All ya gotta do is pint her up thar in the thicket, pull the trigger and go up thar and git him."

On Sunday, since they were snowbound, the loggers held a shooting match. The web of a piece of rail set on a scaffold was their target. Massey's new gun using steel bullets would shoot through the rail, but the bullets of the lighter guns would not penetrate it.

Watching the sport from a distance, the young Carroll had shouted, "You fellows better be careful, one of those bullets will ricochet and hit someone."

"Ricochet, hell, listen to that kid's high-fallutin' talk. Anyway, who ever heard of such a damn-fool thing?" scoffed the owner of a 25-35 as he pulled his trigger. Someone had to remove the bullet from the middle of his forehead with a pocket knife.

Loggers spent much of their time helping the lumber companies build their own railroads in the woods. Leaving the main line of the Alamogordo and Sacramento Mountain Railroad at Russia numerous logging railroads ran down the canyons and along the ridges of the Sacramento Mountains. Some were short, while others were of impressive length. They collected the logs cut in the forest and took them to the Alamogordo and Sacramento Mountain Railway for transportation to the sawmills in the town.

A logging railroad was usually built in the bottom of a canyon. Teams hauled timber in "trains," several logs fastened together by chains, down the mountainsides to the railroad tracks. A distance of a mile, or even two, was not unusual for logs to be pulled on the ground to the waiting cars.

Ways of moving logs[4] to the logging railroads included:

1. Horses pulling strings of logs.
2. A log-slide, a greased V-shaped trough, moving logs to the canyon floor.
3. An aerial tramway carrying logs over rough terrain, canyons, etc. Portable steam boilers operating hoists were used.
4. Steam skidders.
5. Cat logging with a bulldozer and a giant two-wheel cart.

At times construction of the logging railroad spur was not done until after the canyon was logged and logs were piled on the canyon floor. After the logs were cut and moved to the logging railroad, they were loaded on cars by means of a crane. Each car was equipped with a set of rails on top which ran its entire length and met those of the adjoining cars. On these rails the loader could travel from car to car the whole length of the train.

HEISLER #15 on G. E. Breece Lumber road down Peñasco in Sixteen Springs Canyon.
(O. Rasmussen Collection)
Note: This photograph was taken by E. S. Shipp. The location is the Mescalero Indian Reservation, probably in Silver Spring Canyon.

After all the desirable timber had been loaded and removed, workers dismantled the track rails along with usable cross-ties and laid them up another short canyon. New cross-ties were always in demand and many of the men, after they had finished work for the day, earned some extra money by hand-hewing ties.

Builders of the logging railroads, working hastily, had little regard for sharp curves, steep grades or good roadbeds. Their only objective was to get the logs moved. For such rugged country, sturdy and well-built locomotives were necessary. The first locomotive of the Alamogordo Lumber Company was a 3T Shay built for C. M. Carrier of Carrier, Pennsylvania, in April, 1895. Its light weight was 86,200 pounds. (A description of all lumber company locomotives may be found in the Appendix.)

ALAMOGORDO LUMBER COMPANY engine terminal in James Canyon at the mouth of Pumphouse Canyon, ca. 1930–1904.
(Orris Smith Collection)
Note: The Alamogordo Lumber Company engine terminal location is in Russia Canyon, about 2 miles from Russia Station on Alamogordo Lumber Company spur.

Three additional locomotives, all with light weights of 128,200 pounds, and one weighing 137,200 pounds, completed the five original engines. Lima #483 became Alamogordo Lumber Company #1 and, years later, was Southwest Lumber Company #2.

Shay #2 was so powerful that it could pull itself off the rails going up a steep grade with a load. The speed of the engine could drop to a point where it would lose its grip on the rails and spin its wheels; consequently, the wheels climbed the rails and dropped the engine onto the ties. Its crew, going to Agua Chiquita, one Saturday afternoon, was anxious to get to Marcia. On the steepest grade the men should have cut the train but they decided to save time by taking it in one section. Engine #2 jumped the track and the men did not get to the logging camp until after dark.

With steep grades, sharp curves and rough tracks, accidents were so commonplace that crews spent many long hours rerailing locomotives and cars. Rerailing locomotives was especially unpleasant since the driveshafts on the Shays became disconnected and had to be reinserted after the rerailing was completed.

Bridges occasionally gave way dumping cranes and locomotives into the ditch. Fortunately, the loggers avoided bridges as much as possible and preferred instead to make detours around places requiring them. In 1938 a loader went through a bridge in Hubbell Canyon and several days passed before it was back on the track. All traffic stopped until the bridge could be repaired.

From Russia to Marcia, in Peñasco Canyon, the logging company railroad was unusually steep. Most irregular and confusing is evidence of two railroads in that canyon. In places the later road, built on top of the other, has its grade half covering the old cross-ties of the early track. Possibly, the area with its railroad, after having been logged out, was forgotten for some time. When logging started east and south of Marcia by Southwest Lumber Company, the new railroad may have been built. Washouts might also have been the cause for rebuilding of the railroad. For over twenty years parts of an engine lay scattered between Russia and Marcia. In 1921 or 1922, its boiler exploded where the Davis ranch house now stands.[5]

Marcia was a village having, at one time, at least one hundred inhabitants. Engine facilities along with a logging office and commissary made it invaluable to activities in the territory. Mr. T. A. Muirhead managed the commissary and post office. The schoolhouse served as the church on Sundays, as well as a building for all weekly community gatherings, including dances. Dr. Shields, the local country doctor, complete with pill bag, attended all. It made little difference whether they needed a dentist or an obstetrician. He lived in Marcia and traveled horseback or by horse and buggy.

ALAMOGORDO LUMBER CO.
Switchback Nov. 30, 1903. Steel crew Shay #2.

WRECK ON Alamogordo Lumber Co. Railroad about 1902 with Shay #2 sitting across track.

Men employed by the lumber company received the same hospital benefits as railroad employees. A hospital agreement,[6] dated May 29, 1900, stated that an employee must pay:

> 50¢ a month if wages were less than $60.00.
> 75¢ a month if wages were less than $90.00.**
> $1.00 a month if wages were over $90.00.

The hospital agreed to care for all injuries and sickness except venereal disease, smallpox and consumption. It would furnish medicines and dressings. When necessary, the patient would be taken to the Alamogordo Hospital and nursing, board and treatment were free.

In a letter, dated February 27, 1908, by order of the general manager, these hospital benefits ended March 1, 1908.

During the political campaign of 1908, when William H. Taft ran against William Jennings Bryan, feelings ran high at the logging camp in Cox Canyon. By that time, men who had worked since the beginning of operations called themselves "the Eddy crowd" or "the boys." Newcomers discovered it a bit difficult to be accepted. General Manager Tibbets had hired a hotel operator from Indiana, with no timber experience, to become woods superintendent in the mountains. Matters became so serious that the camp doctor, Thomas A. Haxby, fearing bloodshed, called a temporary strike.

Manager Tibbets, expecting the usual thirty carloads a day, asked Haxby, "Why are we not getting any logs?"

"The boys are on strike," Haxby answered. "They'll settle it in a few days."

Arkansas Smith, a widower with three children, agreed to press charges against the superintendent from Indiana. He claimed that the man had whipped his children for playing around his house.

** Presumably this should read 75¢ a month if wages were between $60.00 and $89.00.

LOG TRAIN on S-Bridge, 1910.
(Orris Smith Collection)

MEXICAN CANYON—#2506—shows dog house. Note rails on logging cars behind engine. They are for the log loader moving from car to car, 1936.
(E. Clack Collection)

An Englishman by the name of Powell, one of "the boys," was the Justice of the Peace holding court. In the middle of the trial Judge Wharton, attorney for the prosecution, asked for a court recess. Judge Powell granted the recess and the boys held a short poker game. During the game Wharton said, "Boys, we're going pretty strong. I think we can make a compromise."

Court reconvened during the middle of the afternoon. Wharton submitted a compromise which was agreeable to Judge Powell. A proviso stated that there would be no further proceedings in the case if the superintendent would go back to Indiana. Underneath it all had been his statement concerning the presidential election, that "All the brains in the United States were located north of the Ohio and east of the Mississippi." Logging activities resumed immediately.[7]

Land owned by the lumber company in the Sacramento Mountains contained timber that averaged from eighteen inches to two feet in diameter. Occasionally a log was too large for the band saw at the mill to handle. One log with a diameter of six feet and one inch made ten sixteen-foot saw logs. Red and white spruce, pine and oak were the varieties most used. Thirty percent of the raw material became railroad ties. Mines in Arizona and New Mexico used all the mine timbers that could possibly be produced.

In April, 1918, the Sacramento Mountain Lumber Company bought some of the Alamogordo Lumber Company properties and the mill which burned not long after the transaction.**

The Southwest Lumber Company, controlled by Louis Carr, in July of 1920, bought the remaining timber properties of the Alamogordo Lumber Company and the Sacramento Mountain Lumber Company. Carr rebuilt the mill which ran almost continuously until 1945. The cost of the new mill was $350,000,

---

** The Alamogordo Lumber Company closed down operations in late 1907 and sold most of their remaining assets to Sacramento Mountain Lumber after a 10-year shutdown to resolve government lawsuits on land ownership.

SOUTHWEST LUMBER CO. #6.
(Forest Service, O. Rasmussen Collection)

with an additional $45,000 to build a second mill in the same building. J. W. Kendall, still living in Alamogordo, started as mill foreman for the Southwest Lumber Company in 1920. At times, employing nearly two hundred men, the company cut 125,000 feet of timber a day and, during its existence, produced millions of feet of lumber.

The same year that Carr bought the Alamogordo Lumber Company, the Mescalero tribe entered into a contract with the Cloudcroft Land and Lumber Company for cutting timber on the Elk-Silver Creek unit of the Mescalero Reservation.

Today, the only living member of the directors of the Cloudcroft Land and Lumber Company arrives, each day, at the Cloudcroft Post Office, at 8:30 a.m., for his morning paper. Eighty-four years weigh lightly on Ben Longwell's youthful appearance and retentive memory. His experience working for the Alamogordo Lumber Company as a youth and, later, as a Forest Ranger led to a seemingly promising venture in the logging business.

With his partner, C. M. Pate, in 1920, Longwell organized the Cloudcroft Land and Lumber Company. This company erected several sawmills and a planing mill. In 1923, the company decided to build a logging road from just below the Cloudcroft depot, across the block in front of Main Street, onto the reservation. A most severe winter prevented construction during the first year. Steam shovels could not penetrate the deeply frozen ground.

During that time, Pate was back east handling financial arrangements while Longwell kept his crew in the forest creating a log reserve or delivering logs to each side of the proposed

SOUTHWEST LUMBER CO. #6.
(Forest Service, O. Rasmussen Collection)

LOG TRAIN leaving Cloudcroft, Breece turnoff in foreground.
(Andrew Hendrix, O. Rasmussen Collection)

construction. The workers stacked 5,000,000 feet of timber along the right-of-way. By early spring, part of the crew resumed construction work and finished approximately ten miles of track by July, 1924. The first 8.5 miles of track cost $165,000.

Equipment bought by the company consisted of El Paso and Southwestern engine #185 renumbered #1, and one combination steam shovel and log loader, bought for $14,500 plus $1400 shipping charges. Arrangements included an easy payment plan of $700 a month. For a few weeks the Cloudcroft Land and Lumber Company leased a Shay from the Southwest Lumber Company, while its own locomotive was being overhauled. The company used no other locomotives.

Apparently, the Cloudcroft company constructed its track fairly well, for a logging road, as few accidents occurred. Maximum grade was two percent and the maximum curvature was less than thirty degrees.

The company operated two logging camps. Most of the workers lived in large tents. Rail operations consisted of a daily round-trip with the only engine. Number 1 could handle twelve loads on the return to Cloudcroft. Each car carried 4,400 board feet, while Southwest Lumber Company cars hauled 3,300 board feet on each car.

The Cloudcroft Land and Lumber Company, during its logging operations did about 25,000,000 feet of cutting. After two years, the company took voluntary receivership. W. A. Hawkins, for years a friend of Longwell, as receiver of Cloudcroft Land and Lumber Company, turned over its assets to the Breece Lumber Company. They included the rights to cut and remove timber on the Mescalero Reservation, one log loader and logging equipment, a logging railroad running from Cloudcroft to Silver Spring Canyon, one steam shovel, one water tank car, all property and real estate for right-of-way and other real estate.

LOGGING TRAIN in national forest near Cloudcroft, New Mexico.
(Forest Service, O. Rasmussen Collection)

ALAMOGORDO LUMBER CO. SHAY #5 in Alamogordo yards, 1915.
(Vernon J. Glover, Jr., Collection)

About three years later, Longwell and Pate secured eighteen or twenty million feet of standing timber located twenty miles south of Cloudcroft. They built a railroad, operated by Southwest Lumber Company, from Marcia up Water Canyon. Southwest Lumber Company bought Longwell's timber along with adjoining government timber. Before the operation was completed, Mr. Pate died. His son-in-law, Charles K. Caron, helped finish the transaction.

Noteworthy was another timber business operating at this time and deriving its supply from near Cloudcroft. Ray Daniels had a yearly contract with the American Smelting and Refining Company of El Paso, from 1928 to 1945, to ship aspen poles to the smelter. The poles, put in the molten ore, acted as a flux in one step of fire-refining copper. It is necessary, in order to cast flat shapes, to remove oxygen from metallic copper. This is true for copper to be refined electrolytically and also for copper, less pure, that is used for the manufacture of bearings, bushings and the like.

When Breece Lumber Company bought Cloudcroft Land and Lumber Company, it immediately added twelve miles of track. In the year following, it constructed several additional miles of spur and a sawmill in Elk Canyon. Breece was the only company to utilize fully modern (at that time) methods and equipment, thereby developing the largest logging operation that the area had seen.

At the peak of railroad logging, when both G. E. Breece and Southwest Lumber Companies were operating, it is estimated that nearly 20,000 cars of logs a year were carried down the mountain. Breece Lumber Company alone shipped 11,352 cars of logs during the first year of its operation. During the summer months of that period, the Southern Pacific Railroad ran five trains daily from Alamogordo to Cloudcroft, many of them with two engines. Even in winter, two trains made the trip daily. The Alamogordo Lumber Company had owned 110 log cars. When Breece started operations, the railroad moved in 200 more cars.

Operating south of Russia, the Southwest Lumber Company, however, was often having difficulties. Water was continually a problem because of its high mineral content and muddy settlings. In spite of frequent washings and blowdowns, in which water was released at high pressure from the bottom of the boiler to carry mud out of it, boilers became clogged with scale. Southwest Lumber Company often had two engines at a time in the Marcia shops undergoing scale removal. After flues (pipes) were pulled out of the boiler, a scaler scraped off the hard encrustation before they were put back into service. The heavy scale could cause overheating of the metal in certain areas and create danger of a boiler explosion.[8]

In November of 1936, at Marcia, Tom Wilcox had taken the place of the regular night hostler who had gone hunting. Heisler #3 blew up and its boiler went approximately one hundred feet into the air and across an arroyo near the engine house. Wilcox, also blown into the air, fell through the porch on Ernest Rogers's house. Vernon Petty, manager of Southwest Lumber Company, rushed him to the hospital in Alamogordo where he died the next evening.[9]

The Breece sawmill operated from 1927 until the early 1930s when economic conditions during the Depression caused it to close. Between 1935 and 1940, Breece had small circular mills in the woods which furnished rough green lumber that was shipped to the planing mill in Alamogordo for processing.

Walter B. Gilbert of Albuquerque, in 1939, signed a contract with Breece Lumber Company to dismantle the logging railroad north of Cloudcroft to their main logging camp. Operations, which scrapped and salvaged all of the equipment in the camp, included the old roundhouse and machine shop. There were a number of logging trailers, about ten crawler tractors, ten or fifteen railroad cars, four locomotives and machine shop equipment. Tractors and machine shop equipment were hauled out and sold.** Freight cars and locomotives were cut up and the scrap was disposed of in El Paso.

In the spring of 1940, Gilbert's crew of six to ten men commenced taking up the track, starting at the logging camp, by ripping the rails loose from the ties with an especially built sled apparatus pulled by a tractor. The rails were then loaded onto light cars on the track and pulled by a truck adapted to run on the rails. This worked fine and would have been a good operation except that the old road bed and ties were in such poor condition that the workers spent much more time repairing the track in order to haul out on it than they did taking it up and loading it. Mr. Gilbert recalls that the overall operation, about two years before the war, when the market for scrap was very much depressed, did not show any profit.

Early in 1941, M. R. Prestridge Company acquired the Alamogordo plant, the Mescalero timber contract and fee timber from Breece. This company wanted logs forty feet in length. Log cars on the little railroad could carry logs only twenty feet long around its many curves. For this reason Prestridge Company started using trucks instead of the railroad from Cloudcroft to Alamogordo.

It may be assumed that, for the Southwest Lumber Company, the economic advantage of logging by rail over logging by truck was small, especially in view of the extremely rugged terrain that their railroad traversed. Toward the end of railroad logging, this company had a railroad from Marcia, elevation 8350 feet, in the bottom of Peñasco Canyon, to the bottom of Hay Canyon. To reach this objective, the railroad had to run to the Bluff Spring, elevation 8000 feet, and to climb to the "Summit," elevation 8900 feet, by way of Willie White Canyon and two switchbacks. From the "Summit" it dropped to the floor of Wills Canyon, using three more switchbacks, and continued

** There were five locomotives at the Breece camp. Two (#6 and #15) were hauled out by truck.

down the canyon to one of the company's small sawmills, elevation 8250 feet.[10]

The tracks from the sawmill wound along the mountainsides and canyons to the top of a ridge between Wills and Hay Canyons at an elevation of 8750 feet. From there to the bottom of Hay Canyon the trains ran over what surely must have been one of the wildest stretches of track in the country. Six switchbacks carried the railroad down to an elevation of 7900 feet at the bottom of the canyon. In such terrain operating expenses were naturally high. The coming of the war with its resulting demand for steel was the "coup de grace" for many marginal railroad operations and, apparently, Southwest Lumber's was one of them.[11]

It was in 1941 when the canyons last echoed the rapid exhaust of a geared Heisler that sounded as if it were going sixty when it was actually going only six miles an hour. It was then that Heisler #15 was wrecked in Willie White Canyon where it finally came to a halt wedged in a sharply curved cut.** This, also, hastened the end for the logging railroad.[12]

No train ran over the track between Cloudcroft and Marcia for timber after 1942.** The tracks from the woods to Marcia were taken up that fall and the equipment left at the camp. The end came when a Shay engine was fired up at Marcia in 1945 and sent out over rusty track and weed-choked ties to gather up the idle cars and engines and take them to Russia. When the chores were finished, the Shay's fire was killed. The roar of a logging locomotive, walking up a six percent grade with a

---

** Heisler locomotive #15 was wrecked near the end of the second switchback in Wills Canyon, where a steep recovery track was built to retrieve the engine.

** Railroad logging ended in 1942, but for a time logs were reloaded from trucks to railcars just above Marcia at the mouth of Water Canyon. Thus the railroad from Marcia to Russia apparently was used after 1942. SWLCo. also reloaded logs at Cloudcroft and continued to use the Cloudcroft branch until operations ceased in late 1945.

heavy train, or the triumphal whistle of the engine, as it finally reached the summit, was gone forever from the Sacramento Mountains.

In 1946, the Southern Pacific sent out several trains over the long unused track to Russia to pick up the Southwest Lumber equipment. It was at that time that Engineer P. S. Peterson and Wilbur Fifer, the brakeman, obtained a contract for taking the four 3T Shays, still belonging to Louis Carr, down the mountain to be scrapped. At Russia they made a train, using engine #2510, with empty flat cars between the Shays for braking. At Cloudcroft, Peterson realized that he needed water but had run the whole train past the tank and could not back up to get it. He cut off the engine, chained the Shays, flat cars and caboose to the rails in front of the depot and took the crew to Alamogordo on the engine. Time had not permitted his getting water at Wooten and going back to Cloudcroft.

He realized, that evening, that the risk involved in the project did not warrant a further attempt. Since his opinion as an engineer was highly respected, the Shays were cut up in Cloudcroft.** Even the steel of the logging railroad from Marcia to Russia was ingloriously hauled out on trucks and sold or junked.

** Several reports indicate that the Shay engines were not cut up in Cloudcroft, but were eventually taken down to Alamogordo and shipped to a logging railroad in Mexico (per Fred Bonnell, who was in charge of disposing of the logging railroad assets for M. R. Prestridge).

CHAPTER SIX

Last Stop

THE JOYOUS EXCURSION TRIPS on the little yellow cars had already been terminated on September 20, 1930. Regular trains, once scheduled three times a day, by May 27, 1934, ran only three times a week. The Southwest Lumber Company, controlled by Louis Carr, had purchased all holdings of the Sacramento Mountain Lumber Company as well as the Lodge at Cloudcroft. Primarily interested in timber, the lumber company did not encourage hotel business or passenger service.

Although logging accidents were so common on the tortuous railway that it was often remarked that a report was required stating why not, if there were none, had there never been a serious accident involving a passenger. Even so, suits, developing from slight or imaginary injuries, may have contributed to the abandonment of passenger service on the line. Everything considered, it was to the advantage of the Southern Pacific Railroad Company to haul only freight, no passengers or mail, after February 13, 1938.

Prestridge Lumber Company, hauling timber by truck from the already dismantled Breece holdings north of Cloudcroft, bought Southwest Lumber Company in 1945. Since Prestridge Lumber Company, then, controlled all logging interests in the Sacramento Mountains, the faithful mountain railroad was doomed.

The Interstate Commerce Commission's order for abandonment, on May 7, 1947, gave six months for protests. On September 12, Ernest Clack brought the last scheduled train behind engine #2511 down from the clouds. Two days later the company started pulling track.

The Commercial Construction Company, Incorporated, of Dallas had the contract for dismantling the railroad. The superintendent arrived with engineer, brakeman, foreman for laborers and a caterpillar tractor driver. He conferred with W. R. Adair, Assistant Superintendent of the Southern Pacific, concerning leasing equipment. As a precaution, Adair required that Conductor E. P. McCrary and a crew that he would select be employed if railroad equipment were used.

McCrary's crew included Red Robertson, engineer; L. Mixon, fireman and C. A. Anderson, brakeman. These men, as well as any others who had ever worked for him, had great admiration for "Mr. Mac."

The next morning, with about twenty-five Mexican laborers on trucks, all the men started for Russia. The train crew and visitors were on the work train. When they arrived at the switchback, the Dallas engineer refused to ride farther and walked to the highway. Both he and the Dallas brakeman decided that the mountain railroad was not up to their safety standards.

Starting at Russia with engine #2511, the train began hauling steel to Alamogordo. Before long, the boiler started leaking and it became necessary to use engine #2510. None of the locomotives had been put in good repair and the dismantling project used them strenuously. When #2510 developed air trouble, engine #2506 finally completed the undertaking.[1]

After taking the cars loaded with steel to the wye in Alamogordo and setting them out on its east leg, the men always stayed overnight in the town. The main line from El Paso would have empty cars waiting for them on the west leg. Having been tied up at the wye overnight, before daylight, the train of empties would start up the mountain.

When the track was removed to a point just below the water tower at Cloudcroft, it became necessary to have the cars ahead of the engine and water tank. Otherwise it would have been impossible to fill the water tank. The arrangement going downhill was quite unusual. Since there was no caboose, the five flat cars headed the train, then the scrap car (gondola), the engine, tender, and finally, the water tank followed.

The laborers included spike pullers who put spikes into the scrap car. Others loaded cars with used rails. Some of the rails were sold to Mexico. Many of them were too crooked to be reused and, with the spikes, went to El Paso to the smelter. On the front of the caterpillar was a plow which loosened the ties. Trucks hauled them out and ranchers bought many for fence-posts. Some of the trestles remained standing but local buyers have razed and sold the timbers from others. Several residents in the area have used them for beams in ceilings.

On one of the trestles the laborers were pulling spikes. With all his strength one man tried to remove a spike, but it did not let go. He tried a second time, not realizing he had already loosened it, and it gave way instantly. The man, the spike and his bar all went over the bridge. He, first, hit a stringer, which broke his fall, then he fell on to the ground, altogether a distance of nearly fifty feet. Anderson rushed him to the hospital in Alamogordo and X-rays showed the only damage to be some cracked ribs.

As long as the crew went as far as Wooten, they often took apples home for pies. Along with coffee that had brewed all day on the boiler head of the engine, homemade pie never tasted better.

During one trip, McCrary and Anderson were in the dog house, a little room on top of the tender for one man to protect himself from the weather. McCrary had made himself comfortable and had even removed his false teeth. Suddenly, Anderson saw a rock about three feet in diameter in the middle of the track. He shouted, "Pull the air, Mac! Pull the air!"

THE LAST REGULAR TRAIN pauses for its portrait on S-Bridge, Sept. 12, 1947.
(Last photo by Dan Kelly. Courtesy of E. Clack)

(Courtesy of E. Clack)

Fumbling in his pocket for his teeth with one hand and under the seat for the brake valve with the other, "Mr. Mac" did not get the train stopped until after it had hit the rock.

The switchback was the most difficult place on the line for the train with the carloads of steel. Because of the steep grade, the cars would run out of air below it and the train would run away nearly every day. Eventually the crew would restore the air and be able to stop the train.

Not all problems on this piece of work happened in the mountains. The train was at the wye in Alamogordo where one man had been instructed to pipe retainer air on the cars to enable the engineer to control the train going down the steep grades. Anderson coupled the engine onto a half dozen cars. He had knocked all the brakes off, except on the last car, the night before. When he pumped up the air the next morning, he found one car with a bad triple valve which would have made the brake inoperative. He took that empty car out and set it against the loaded cars. Then he went back to get the empties but found that the man had let the handbrake off and all his cars had rolled down to the main line. They hit the derail and went off the track onto the ground. Train #43 was coming and for its crew the confusion of signals presented brief unnecessary consternation.

Anderson helped get the cars back on the track and started thinking of the consequences. He discussed the matter with "Mr. Mac" concluding with, "If that boss even bawls me out, I'll quit."

McCrary, getting a bit tired of the whole project also, answered, "If you quit, I'll quit, too."

When Anderson met the construction company head, he explained the situation to him. The man's only remark was, "I'll be damned," as he turned and walked away.

Anderson reports, "We didn't get to quit," then adds with pride, "and with what I made on that job, I paid for our daughter Georgia Ann, who is now a senior at Ysleta High School."

Heartlessly dismantled within six weeks, by September 1, 1948, was the courageous railroad into the mountains, which, for so long, had furnished pleasure or prosperity for so many. Old-timers like to imagine that it just climbed a last cloud and, with it, floated away.

APPENDIX

From the files of Vernon J. Glover, Jr., Albuquerque,
New Mexico. Editor of *New Mexico Railroader.*

The following standard form rosters may best be explained by using engine # 101 as an example, thus:

101 2-8-2T Baldwin # 16103, 7 / 1898 46-21x24-135000-160-31290

101 is the road number, the number assigned to the locomotive by the operating company.

2-8-2T represents locomotive wheel arrangement. First digit counts pilot truck wheels. Second digit counts drive wheels. Third digit counts trailing wheels. Zero is used if there are none. The "T" means a tank locomotive carrying fuel and water on locomotive rather than on separate tender. Geared locomotives are represented by 3T Shay or 3T Heisler, 3T being number of driving trucks under engine plus name giving type of locomotive.

Baldwin, #16103, 7/1898 gives builder, builder's serial number and date of construction.

46-21x24-135000-160-31290 gives dimensions and weights of locomotive.

46 is diameter of drive wheels in inches.

21x24 is diameter and stroke of cylinder in inches.

135,000 is total engine weight in pounds. Sometimes preceded by a lesser figure representing weight on drive wheels only.

160 is the working steam pressure in pounds per square inch.

31290 is tractive effort (pulling power) of locomotive in pounds.

LOCOMOTIVES IN THE SACRAMENTO MOUNTAINS OF NEW MEXICO
1. Main line from Alamogordo to Cloudcroft and Russia

ALAMOGORDO & SACRAMENTO MOUNTAIN RY.

Numbers		Descriptions	
101	2-8-2T	Baldwin #16103, 7/1898	46-21x24-135000-160-31290
102	2-4-2T	Baldwin #13361, 4/1893	44-14x24-72130-46630-140-12722
103	2-8-0	Baldwin #16494, 2/1899	46-21x24-140600-124800-160-31290

104	2-8-0	Baldwin #17107, 10/1899	ditto
105	4TShay	Lima #673, 3/1902	40-(3) 15x17-260300-180-43605 Lima Class 125-4
99	4TShay	Lima #1893, 8/1907	46-(3) 17x18-357300-200-59116 Lima Class 150-4

EL PASO & SOUTHWESTERN RR.
SOUTHERN PACIFIC LINES

EP&NE	EP&SW	SP			
52	181	2505	2-8-0	Baldwin #17397, 3/1900	50-21x26-141000-126000-160-31190
53	182	2506	2-8-0	Baldwin #17398, 3/1900	ditto
54	183	2507	2-8-0	Baldwin #17443, 3/1900	ditto
55	184	2508	2-8-0	Baldwin #17444, 3/1900	ditto
A&NM	EP&SW	SP			
19	217	2510	2-8-0	Baldwin #20237, 3/1902	51-21½x28-176000-160000-180-38830
24	218	2511	2-8-0	Baldwin #26656, 10/1905	ditto

LOCOMOTIVES IN THE SACRAMENTO MOUNTAINS OF NEW MEXICO
Histories

101. Built for New Mexico Coal and Railway Co. as A&SM 101, to EP&NE 101, EP&SW 101, 201, 401, 420, rebuilt to 0-8-0 switcher, to SP 1300 Class SE-1, scrapped July 24, 1934.
102. Built for Baldwin Locomotive Works exhibit at 1893 Chicago World's Fair, to New Mexico Coal and Railway Co. 6 in Dec., 1893, to A&SM 102, sold to EP&SW 2/1903, to EP&SW 202, 402, 2, to F. C. Nacozari 25 in 1906, to United Sugar Co. at Los Mochis, Sinaloa, in 1921, returned to F. C. Nacozari in 1926, scrapped Feb. 1, 1935. A&SM 102 was sold to EP&SW before the EP&NE was purchased by the EP&SW, therefore the 102 never was owned by EP&NE.
103. Built as A&SM 103, to EP&NE 103, to EP&SW 185, to Cloudcroft Lumber & Land Co. 1 on April 24, 1924, to George E. Breece Lumber Co. 1 in 1926.
104. Built as A&SM 104, to EP&SW 186, to SP 2504, Class C-14, scrapped Sept. 27, 1935.
105. Built as EP&NE 105, to EP&SW 100, to F. C. Mexicano 110 in 1905, out of service by 1923.
99. Built as Norfolk & Western 56, to EP&SW in 6/1916, to Red River Lumber Co. of California in 1920, scrapped in April, 1929.

During the later years of the A&SM line various of the smaller 2-8-0s of the El Paso & Southwestern and Southern Pacific were used. The locomotives commonly seen on the line are listed with their various numbers over the years.

During the 1940s, Southern Pacific 2507, 2510, and 2511 were assigned to the line, and 2511 hauled the last train in 1947.

SP 2505 through 2508 were built for the EP&NE and SP 2510 and 2511 were built for the Arizona & New Mexico Ry., a copper hauler running from Lordsburg, NM to Clifton, AZ, and purchased by the EP&SW in 1921.

2. Logging lines of ALAMOGORDO LUMBER COMPANY, 1898–1918
SACRAMENTO MOUNTAIN LUMBER COMPANY, 1918–1921
SOUTHWEST LUMBER COMPANY, 1921–1942**

Numbers			
ALCo	SMLCo	SWLCo	Descriptions
1	1	2	3T Shay Lima #483, 4/1895 32-(3)12x12- 86200# light wt.
2	2	3	3T Shay Lima #568, 3/1899 33-(3)14x14- 128200# light wt. Lima Class 65-3
3	3	4	3T Shay Lima #580, 9/1899 33-(3) 14x14- Lima Class 65-3
4	4	—	3T Shay Lima #587, 11/1899 33-(3) 14x14- Lima Class 65-3
5	5	5	3T Shay Lima #700, 6/1902 36-(3) 14x14- 137200# light wt. Lima Class 65-3
—	—	1	3T Shay Lima #3155, 6/1921 36-(3) 12x15- Lima Class 70-3
—	—	3	3T Heisler Heisler #1540, 2/1927, 70 ton, 3-truck
—	—	15	3T Heisler Heisler #1534, 1926, 70 ton, 3-truck
—	—	6	3T Shay see notes below

LOCOMOTIVES IN THE SACRAMENTO MOUNTAINS OF NEW MEXICO

Histories

Lima #483.	Built for C. M. Carrier, Carrier, Pa. Then to Alamogordo Lbr. Co. May have been renumbered 6 in later years on Southwest Lbr. Scrapped.
Lima #568.	Built for F. L. Peck, Scranton, Pa. Then to Alamogordo Lbr. Co. Scrapped.
Lima #580.	Built for Alamogordo Lbr. Co. Scrapped.
Lima #587.	Built for Alamogordo Lbr. Co. Wrecked by boiler explosion 1921 or 1922.

** Logging locomotives of the ALCo, SMLCo, and SWLCo. No photograph of an Alamogordo Lumber Company engine #1 has ever been found; instead, a photograph shows ALCo engine #6. It appears that Lima 483 was numbered 6 on ALCo, SMLCo, and SWLCo.

Lima #700.	Built for Alamogordo Lbr. Co. Scrapped.
Lima #3155.	Built for Southwest Lbr. Co. Scrapped.
Heisler #1540.	Built for Southwest Lbr. Co. Wrecked by boiler explosion late 1938.
Heisler #1534.	Purchased 1940 from George E. Breece Lbr. Co. Scrapped.
Shay #6.	Probably was Lima #483 renumbered in later years by Southwest Lbr. Co.

Locomotives in service in 1937:		
	1	3T Shay (Lima #3155)
	2	3T Shay
	3	3T Heisler
	5	3T Shay
	6	3T Shay

3. Logging lines of CLOUDCROFT LUMBER AND LAND CO., 1924–1926
GEORGE E. BREECE LUMBER CO., 1926–1940

Numbers		Descriptions
1	2-8-0	Baldwin #16494, 2/1899 46-21x24-140600-124800-160-31290
3?	2T Shay	Lima #2611, 1/1913 36-(3) 10x10-
6	2-6-2T	Porter #6727, 8/1922 44-19x24
15	3T Heisler	Heisler #1534, 1926 70 ton, 3-truck
102	2-6-2	Schenectady #5626, 8/1900 51-19x24-132000

Histories

1. Built as Alamogordo & Sacramento Mountain Railroad 103, to El Paso & Northeastern 103, El Paso & Southwestern 185, to Cloudcroft Lumber & Land Co. 1 on April 24, 1924, to Breece 1, 1926.
3. Built for Grayling Lbr. Co. 3, Arkansas City, Ark., to O. S. Hawer Lbr. Co., Monroe, LA, to Breece during 1920s.
6. Built for McKinley Land & Lumber Co., a Breece predecessor at Grants, N.M. Transferred to Cloudcroft during 1920s. Trucked out 1940, disposition not certain.
15. Built for Breece. Sold 1940 to Southwest Lumber Co.
102. Built for Colorado Springs & Cripple Creek District RR 102.

ACKNOWLEDGEMENTS:

Baldwin-Lima-Hamilton Corp.
P. E. Percy
Ralph D. Ranger, Jr.
Bert Ward
Joe Strapac
Gerald M. Best
Joseph Webber
Walter Casler
Olaf Rasmussen
Allen Copeland

REFERENCES

CHAPTER I

1. Interview with Mrs. A. T. Seymour (Betty Hawkins), Albuquerque, New Mexico, July 15, 1964.
2. Interview with Carroll Woods, Alamogordo, New Mexico, July 18, 1964.
3. Woods.
4. William A. Keleher, *The Fabulous Frontier* (Albuquerque: University of New Mexico Press, 1962) p. 281.
5. Woods.
6. *Ibid.*
7. Mrs. Tom Charles, *Tales of the Tularosa* (Alamogordo 1954) p. 42.
8. Emily Kalled Lovell, *A Personalized History of Otero County, New Mexico* (Alamogordo: Star Publishing Company, 1963) p. 7.

CHAPTER II

1. Carlsbad *Eddy Current* Vol. VII, No. 9.
2. Keleher, p. 192 (footnote 11)
3. Keleher, p. 257.
4. Interview with S. A. Ramsdale, El Paso, Texas, May 7, 1964.
5. Plat of Location Survey from Alamogordo Junction to Toboggan Gulch, Dona Ana County, New Mexico, October 12, 1898.
6. *The El Paso Times,* August 11, 1898.
7. *History of New Mexico, Its Resources and People* (Los Angeles: Pacific States Publishing Company, 1907) Vol. II, p. 905.
8. Interview with J. D. Hackett, El Paso, October 29, 1965.

9. Interview with Wilbur Fifer, El Paso, October 10, 1965.
10. Interview with Jentry Kendall, El Paso, October 20, 1965.
11. Records of Southern Pacific Railroad Company, El Paso.
12. Henry R. Mulchahey, Assistant Engineer, Southern Pacific, El Paso.
13. Interview with Ben Longwell, Cloudcroft, New Mexico, August 25, 1965.
14. *The El Paso Times,* September 29, 1925.
15. Southern Pacific files.
16. Interview of Mrs. J. O. Frilick to Frank Parsons, Cloudcroft, June, 10, 1964.
17. Southern Pacific files.

CHAPTER III

1. Seymour, July 25, 1965.
2. Mardee Belding de Wetter, "Revolutionary El Paso: 1910–1917," *Password,* El Paso County Historical Society, Vol. III, No. 4 (October, 1958) p. 151.
3. Telephone conversation with Mrs. J. Burges Perrenot, El Paso, February 8, 1966.
4. *White Oaks Eagle,* June 15, 1899.
5. Mulchahey (estimates).
6. Interview with Buddy Ritter, Cloudcroft, August 25, 1965.
7. Estelle Goodman Levy, "The Cloudcroft Baby Sanatorium," *Password,* El Paso County Historical Society, Vol. VII, No. 4 (Fall, 1962) pp. 137–141.
8. *Ibid.*
9. Southern Pacific files.

CHAPTER IV

1. *White Oaks Eagle,* January 4, 1899.
2. Interview with Olaf Rasmussen, Alamogordo, New Mexico, August 28, 1965.
3. *Otero County News,* August 7, 1914.

4. Woods.
5. Fifer.
6. Interview with Edmund Given, El Paso, February 9, 1966.
7. *The El Paso Times,* August 31, 1907.
8. Interview with Mrs. Jenny Dolan, Tucumcari, New Mexico, November 4, 1965.

CHAPTER V

1. *History of New Mexico, Its Resources and People.* Vol. II, p. 905.
2. Southern Pacific files.
3. Woods.
4. Rasmussen.
5. *Ibid.*
6. Southern Pacific files.
7. Interview with Ben Longwell, Cloudcroft, New Mexico, August 25, 1965.
8. Rasmussen.
9. Interview with Mrs. Bertha Wilcox and Mrs. Wilma Hash, Alamogordo, September 15, 1965.
10. Rasmussen.
11. *Ibid.*
12. *Ibid.*

CHAPTER VI

1. Telephone conversation with C. A. Anderson, El Paso, Texas, November 8, 1965.

ACKNOWLEDGEMENTS

Unable to claim either the idea or the material for this history, it is with gratitude and pride that I view its accomplishment. During the holiday season of 1963, Fran and Frank Parsons of El Paso informed me that we should write a book about the railroad that climbed to Cloudcroft. Frank started gathering pictures, taped an interview with Mr. and Mrs. Frilick of Cloudcroft and presented me with an eight-page blueprint of the railroad that reached from my living room sofa to the kitchen sink. Fran typed the first attempt. Eventually this collaboration dwindled, but I am most grateful for all their efforts and inspiration.

Bob Staggs, my son-in-law, of Dallas, was ready from the beginning to do some drawings and Bill Calhoun in the English Department of University of Texas at El Paso pulled things together just before publication. To them both I am exceedingly thankful.

Sources of material were almost overwhelming. Henry Mulchahey, Assistant Engineer of the Southern Pacific in El Paso, furnished records, drawings and pictures, as well as his own inexhaustible information. Always as close as my telephone were his patient answers to my many, many questions. Other available sources of technical advice were Olaf Rasmussen, a senior at New Mexico State University at Las Cruces and Vernon J. Glover of Albuquerque. Olaf has walked over and mapped all the known logging railroad beds in the Sacramentos. Vernon furnished the locomotive roster and criticism on Chapter IV.

Throughout the story it is apparent that vivid accounts flowed from the memories of Betty Hawkins Seymour of Albuquerque, S. A. Ramsdale and Tom Bell of El Paso, Orris Smith of Capitan, Carroll Woods of Alamogordo, and Ben Longwell and Buddy Ritter of Cloudcroft. To these people my appreciation can never be fully expressed.

Others who so kindly furnished detailed information were J. D. Hackett and C. A. Anderson of El Paso, Walter B. Gilbert of Albuquerque, and Mr. and Mrs. Wiley Smith of High Rolls. The Smiths, Ernest Clack of El Paso and Emily Lovell, formerly of Alamogordo, were among those most generous with pictures.

Mr. and Mrs. Ernest Rees and J. W. Kendall of Alamogordo all furnished material for which I am deeply indebted. History of La Luz was graciously furnished by J. J. Gutierrez and Juan García, Jr.

Carl Hertzog's interest, guidance, encouragement and advice on typography have been of inestimable value.

It would be impossible to thank at any length each contributor and unforgivable to omit a single one. Therefore, please accept my sincerest thanks, if you shared in any way the compiling of this "personal landscape."

DOROTHY JENSEN NEAL

INDEX